Digination

The Fairleigh Dickinson University Press Series In Communication Studies

General Editor: Gary Radford
Department of Communication Studies
Fairleigh Dickinson University, Madison, New Jersey

The Fairleigh Dickinson University Press Series in Communication Studies publishes scholarly works in communication theory, practice, history, and culture.

RECENT PUBLICATIONS IN COMMUNICATION STUDIES

Aritz, Jolanta and Robyn C. Walker, *Discourse Perspectives on Organizational Communication* (2011)

Groom, S. Alyssa and Fritz, J. M. H, *Communication Ethics and Crisis: Negotiating Differences in Public and Private Spheres* (2011)

MacDougall. Robert C., *Digination: Identity, Organization, and Public Life in the Age of Small Digital Devices and Big Digital Domains* (2011)

Eicher-Catt, Deborah and Isaac E. Catt (eds.), *Communicology: The New Science of Embodied Discourse* (2010)

Cassino, Dan and Yesamin Besen-Cassino, *Consuming Politics: Jon Stewart, Branding, and the Youth Vote in America* (2009)

On the Web at http://www.fdu.edu/fdupress

Digination

Identity, Organization, and Public Life in the Age of Small Digital Devices and Big Digital Domains

Robert C. MacDougall

FAIRLEIGH DICKINSON UNIVERSITY PRESS
Madison • Teaneck

161 0208

AUG 3 0 2012

Published by Fairleigh Dickinson University Press
Co-published with The Rowman & Littlefield Publishing Group, Inc.
4501 Forbes Boulevard, Suite 200, Lanham, Maryland 20706
www.rowmanlittlefield.com

Estover Road, Plymouth PL6 7PY, United Kingdom

British Library Cataloguing in Publication Information Available

Library of Congress Cataloging-in-Publication Data

MacDougall, Robert C.
    Digination : identity, organization, and public life in the age of small digital devices
and big digital domains / Robert C. MacDougall.
        p. cm.
    Includes bibliographical references and index.
    ISBN 978-1-61147-439-8 (cloth : alk. paper)—ISBN 978-1-61147-440-4 (ebook)
    1. Information technology—Social aspects.   2. Digital media—Social aspects.
3. Mass media—Technological innovations.   4. Telecommunication—Social aspects.
5. Information society.   I. Title.
    HM851.M325 2011
    303.48'33—dc23   2011030833

∞ ™ The paper used in this publication meets the minimum requirements of American
National Standard for Information Sciences—Permanence of Paper for Printed Library
Materials, ANSI/NISO Z39.48-1992.

Printed in the United States of America

To my son, Owen, the new man on campus.
Cave ab homine unius libri.

Contents

Acknowledgments

I must first thank my editor, Gary Radford, at FDU Press for keeping me on time and for his keen suggestions throughout the process. Thanks also to Brooke Bascietto, associate editor, and Liliana Koebke, editorial assistant, at Rowman & Littlefield. Your patience and gentle prodding were just what the doctor ordered. The core of most of the media analyses that follow are the comments, observations, and noticings of dozens of students, colleagues, and passersby who trusted me with their information and who were often gracious enough to allow me to check in again (once, twice, or thrice) to see what was new in their media worlds.

And of course this book would have never come to fruition were it not for the insights, criticism, questions, and answers provided by so many colleagues, friends, and acquaintances also grappling with what it means to live, play, and work in Digination today. These are folks who also offered various sorts of writing advice, engaged me in conversations that sparked new ideas, or just gave me some needed inspiration or moral support. They include Fritjof Capra, Frank Dance, Thom Gencarelli, Eric Goodman, Bruce Gronbeck, Marco Haber, Ronan Hallowell, Don Ihde, Bob Logan, Eric McLuhan, John Orfan, David Paterno, Greg Payne, Valerie Peterson, Tom Porter, Neil Postman, Phil Rose, Leonard Shlain, Tracey Stark, Lance Strate, and Edward West.

The media ecology listserve has been and continues to be a touchstone for me, a virtual third place where I just love to lurk—and periodically argue and debate and share ideas, news, jokes, hopes, worries, and dreams. No other list comes close to the MEA at ibiblio in this regard. I look forward to "meeting" my friends there—those I know, and those I have not yet had the pleasure to meet in person. The MEA, in all its personages and spirits, continues to provide me a great deal of intellectual sustenance and spiritual solace. Every book is a collaborative effort, certainly behind the scenes, and out front as well. But a book like this had to be a community project. I thank all my friends in the MEA for helping this one along. The wisdom, the wit, the sense of goodwill and community—I dare say you'll ne'er find a comparable collection of wide open minds elsewhere in the *nation*.

To my two sons, Charlie and Owen: you guys teach me something new every day about myself, about you, and about this nutty world we live in. Finally, I must acknowledge my beautiful and whip-smart wife, Jennifer. I brought along an image of a blog tetrad to our first date. You humored me that night, bebe, and put a spell on me too. Thanks for your patience, your love, your kindness, and your tireless ear in hearing me out, then and now.

Chapter 1

Understanding Our Digination

"When we try to pick out anything by itself, we find it hitched to everything else in the universe."

—John Muir

INTRODUCTION

This book is about a few old and many new forms of communication. It is about communication technology in general and digital communication systems and applications in particular. I recognize that there are plenty of people around today who do not own a mobile phone, a computer, or even a television. And I know that some of those folks probably think none of this applies to them. But, as I'll argue in the pages that follow, more than any other form of technology perhaps (including the automobile, roads, and electrical grids—or even more conceptual forms of technology like modern theories of education, economics, and medicine) communication media play a particularly powerful and systemic role in our culture, or any culture for that matter.

The argument here is that the introduction or removal of any means of communication tends to create reverberations throughout the entire cultural system. Despite their protestations to the contrary, I'm pretty sure late adopters and even abstainers are not exempt from the psychological, social, and cultural effects and side effects of modern communication technology. While there are certainly varying degrees of immersion—that is to say, while some of us live in the high-rise downtown district, some at the city limits, and still others out in the proverbial "woods"—we all live in Digination today.

Digination is a neologism from the compounded noun *digital-nation*, and is fast becoming a truism. As my old school in the heart of Boston is finally filling in the last of the wireless dead spots across campus, and as the city's mayor continues to promise a fully Wi-Fi'd city "in the next five years," we are quickly replacing (and in the processes quickening the pace of obsolescence of) our analog and other non-digital infrastructures. This is the non-human edifice: the smart buildings, the fiber optics, the high-band radio transmitters—all those quickly forgotten and often invisible things that comprise the physical substance of digitally enabled society. At a surface level, this is a covering, a blanketing, and embedding of the physical environment with the underlying (and overarching) communication system known as the Internet.

Much of this infrastructure is surreptitiously threaded along the old analog lines, buried underground, beamed across open expanses of terrain, and down to and around the landscape from geosynchronous orbit. And we have a sense of what this might look like. It is dendritic and circulatory in both form and function. Cities like Philadelphia, Boston, Atlanta, New York, San Francisco, and Seattle are leading the way domestically. Other cities and city-states around the world are already much more immersed in the digital than the coastal rims of the United States, but the pace of transformation has been picking up on this side of the Atlantic over the past decade especially. One might even call it a feverish pace.

So Digination describes a set of policies, procedures, and processes designed to keep a physical infrastructure running smoothly. These policies, procedures, and processes have been advancing quickly and efficiently in a way most federal and large-scale private infrastructure projects rarely do. Of course, this is understandable, yet it is not uncommon at national conferences or on any number of discussion lists concerning digital technology for people to decry the fact that even in third-world contexts we are seeing things like Wi-Fi and mobile phone infrastructure installed before sewage systems or fresh water conduits. Again, this is understandable and even predictable because the latter requires massive capital investments and a whole lot of time and labor. In contrast, I have a friend who installs wireless networks for a living and he and one assistant can thread the lines, configure the routers, and tune the signals for the equivalent of a small village with mobile telephony and broadband Internet access over a short weekend. Digital infrastructure, in other words, is incredibly cheap in comparison to just about any other utility. Traditional infrastructure conglomerates like GE, ABB, and Hitachi are still doing their thing, but Microsoft, Google, Oracle, Apple, Cisco, AT&T and Verizon and the like are the new infrastructure companies worth keeping an eye on. Of course, this is not to say that digital infrastructures are more valuable or crucial to the social and physical health and overall sustainability of a community—they just propagate more cheaply and more efficiently. How we learn to balance concerns for efficiency and cost-savings of various sorts will be the challenge of the century as rural parts of the first world are incorporated, and as China and India and most of Africa get online in their continued efforts to modernize.

And yet if what's been described so far is all this book were about, I'd stop here and politely direct readers to Anthony Wilhelm's *Digital Nation: toward an inclusive information society* (2004) since, as Wilhelm argues, the digital worlds we are building will enable all kinds of wonderful futures, just as long as everyone has access to it and uses its powerful tools in the right ways. Wilhelm sticks to his guns throughout his book to a time-honored trope: that technology is not the problem, rather it's *how* people use technology that's the problem. I'll argue, more in line with Marshall McLuhan, Neil Postman, and other media ecologists, why it's not so much how we use certain tools that matters, it's *that* we use them.

SOME PERSONAL PROLOGUE

My own excursions into the digital world began in earnest in the waning days of 1980 when, after my younger brother and I had apparently harassed my parents long enough prior to the holidays, we awoke to find a brand new Commodore Vic20 tucked neatly under the tree on Christmas day that year. I did not know it at the time, but that moment landed smack dab in the middle of a couple years' worth of some very big media-happenings. The Vic20 arrived just six months after Ted Turner's Cable News Network aired its first broadcast (June 1, 1980), and about six months before MTV's first music video flashed across television screens. It was, appropriately, the *Buggles'* "Video Killed the Radio Star" (airing August 1, 1981).

About 20 months after the computer arrived, the premier issue of *USA Today*, the first newspaper to mimic the image-based format of the television, hit newsstands (September 15, 1982). It is also during this time (almost exactly one week after the computer materialized under our tree) that Marshall McLuhan, a name I probably did not hear and certainly would not have recognized for another twelve or thirteen years, was finally overcome by the deleterious effects of a stroke. He died in his sleep on New Year's Eve, 1980. Again, if I had no clue about the man or his ideas at the time, McLuhan would come to mean a lot to me in my doings with and thinking about communication technology in the years to come. To suggest how and why, let me get back to that first computer.

Advertised by William Shatner as "the first honest-to-goodness full color computer you can buy for only $299," my father managed to find a Vic20 about a year after its release into the consumer market on sale at Sears for $189. There was precious little my younger brother and I could do with this wonderful machine. He mostly watched while I wrote a few basic programs that created stick-figures and allowed 3-D renderings of simple geometric shapes to appear and bounce around on the screen. For several months, these little programs would be lost to oblivion every time I shut the machine down, but I was eventually able to save my work to the cassette tape drive memory system we picked up a few months later. Being able to save those primordial programs quickly led to a much more intimate connection between me and my machine.

What really made that computer worthwhile at the time was the awesome collection of games we could plug into the cartridge slot on the back side of the all-in-one white plastic keyboard/CPU case. I had games with titles like *Alien Blitz, Buck Rogers: Planet of Zoom, Dungeon Master, Gorf, Lode Runner, Omega Race,* and *Robotron 2084.* And we either bought, traded, or borrowed the requisite Vic versions of the legendary arcade games like *Asteroids, Centipede,* and *Donkey Kong* from the one or two other kids we knew at school with Commodore machines. I also obtained a few games and apps that were particularly enjoyable when played alone. There was the *Vic Biorhythms* program which allowed me to finally understand why I was, presumably, sometimes melancholic on certain days and in superior moods on others. The program was a gimmick of course. It was based, if I recall, on some very arbitrary calculations involving date and time of birth, favorite color, and a few other disparate measurement criteria. But this was beside the point since the program allowed me to at least *feel* as though I had access to something few others did. If a sort of placebo effect was in full effect, I could, in a sense, control certain aspects of my life and, with the proper data, predict things about the lives of my friends and associates. I was twelve, maybe thirteen after all.

Our bedroom soon became a thriving hub of activity with neighboring kids all waiting their turn to play for free what cost a quarter, or two, at the shopping mall arcade just a couple clicks up the main artery in town. Aside from the three or four other Commodore owners I knew around town, only two other associates had this sort of thing at their disposal: the rich kid who lived in the gated community adjacent to our secluded little dead-end street, and my best friend a few blocks away, whose dad was a math professor at the local University. Then again, that was one of those weird Apples that didn't have much for games. My buddy's dad, the professor, was always writing in that French-sounding language called Pascal, computing statistics from the ski jump competitions he was involved with at the Olympic training facility near Lake Placid.

He'd print out these long strings of numbers and tape them to the wall of their dining room. Once I noticed a particularly massive string of numbers hanging there—or, rather, one really long number. This was in fact the first "google" (actually, googol) I'd come in contact with. The rich kid, on the other hand, whose dad owned a Pontiac dealership, had accumulated the complete Atari game collection at the time. And it was all connected to the largest television set I had ever stood in front of. I could have climbed into the damn thing with two or three friends if the screen wasn't in the way. Boy, those were the days—we were as high-tech as kids came.

However the most incredible thing my computer allowed me to do was play a text-based game none of my friends much liked at the time. It was called *Adventure Land*, and if by today's standards it was the most rudimentary Role Playing Game (RPG) and "active plot manipulator," it was the coolest thing to play in the middle of the night when the lights were out. I have fond memories of those phosphorescent green characters pixilating on the screen as I typed my commands—followed immediately by the computer's reply—and the mystery looming around each corner. Without having a clue I was, in effect, tightly involved in a commercial manifestation of Alan Turing's famous test—and man I was hooked. It all seemed so real to me. I'll never forget that first scene description. The game always started out at an amusement park in Sandusky, Ohio. I've never been to Sandusky, at least not in real life, but I'd been there many times in my computer-enabled imagination. I'm not sure if Sandusky ever even had any worthwhile amusement parks, but I was going to a really good one in those wee hours—and I often had no idea who I would meet, nor where I'd end up. Now, it so happens that I got good enough at the game to exhaust all the permutations of people, places, and events, though for a while it was my own private universe; quite infinite in its look and feel. It was in fact a very finite, closed world I was roaming around in, and so my attention started to wander, and I started to wonder if there was more that was possible in this new life in the screen. Right around that time is when *Time* magazine featured the home computer in their traditional "person of the year" spot on the front cover. Something was going on in our culture that was hard to put your finger on.

My friend's dad, the college math teacher, would sometimes spend all day running different calculations on ski jump results or writing "tiling programs" that figured out the best way to fill a truck with oblong milk crates just for the fun of it. It was nothing that interested me, but the man was transfixed—a true absent-minded professor. He was hooked into that machine even worse than I was with my Vic. And that rich kid? He got a fancy IBM "PC," and soon had the first "pornographic" computer program in the neighborhood. It made black and white calendars featuring hazy images of naked girls out of alpha numeric characters that ran off a track-fed dot matrix printer. Still, they were some of the most mystical things we'd ever seen. I would have had trouble explaining precisely why at the time, however they were much more impressive than some old Playboy magazine we might come across on the trails through the back woods, or tossed in the alley behind the local hangouts downtown.

I eventually grew tired of the Vic 20 and all those games toward the end of high school. But brand loyalty is a funny thing, because just a few years beyond that, as a sophomore in college, I picked up the latest machine Commodore had to offer, the sleek 128D ("D" for 5 ¼ inch floppy disk drive). At this point (around 1986) IBM *PC*s (personal computers) were all over the place, early Microsoft GUIs (graphic user interface) were already colonizing desktops, and Apple was making great strides with the MAC, so I'm not sure if it was in fact brand-loyalty or just laziness—or maybe it was because those old Vic game cartridges were compatible, can't really say for sure. I just liked the look of the 128. It had a decent 80-column word processor that I wrote most of my college papers and essays on over the next few years. I'd also write my master's thesis on the 128—spewing out those tractor-fed pages that I had to carefully rip the edges from before final submission. Then, in 1989, I purchased a used 300 baud modem that mated to my machine. But here again, I didn't really know what to do with the new hardware. The 128 was still a leader for gaming at the time. That was always its real niche. The notion of calling in to connect with other systems or other machines seemed unnecessary.

I still wasn't much of a programmer—ten years in, still just copying canned machine language recipes out of the backs of magazines like *Byte* and *Compute!* The little gray modem sat on a shelf in my college apartment gathering dust. I think it was two years later, after I enrolled in graduate school at the State University of New York at Albany, that I learned about the dial-up service. With a little tinkering, that 300 baud modem allowed me DOS-based e-mail access to the IBM/Unix mainframe. If the geologically-slow transfer rate often got bogged down even in the face of my hunt and peck typing style, the day I heard that crackling-buzz followed by the tell-tale high-pitched whine marked my real entrance into the digital world. I was now a bona-fide member of *Digination.*

DIGINATION

At bottom, the process of Digination points to the culture-wide production, consumption, and idiosyncratic interjections of symbolic content ultimately based on zeroes and ones. And this second layer of meaning itself has a dual aspect. The first aspect aligns directly with the process of *digitization*—a verb this time, that describes the action of putting something (words, music, pictures, data) into digital form. The second aspect refers to modes of thought and action that manifest in the alteration of experiential and behavioral patterns happening all around us that are due, in large measure, to the process of digital encoding. There's no question, we now have Digination functioning more like a verb, and I do liken Digination to a kind of socialization or acculturation. This includes any and all qualitative and quantitative changes in relationships and social interactions that are prompted and enabled by these new technologies—things like personal and corporate web pages, blogs, multi-function phones, music players, and communication appliances, laptops and hand-held computers, telecommuting and teleconferencing apparatuses, gaming systems, and all that.

However, all of that still may not distinguish Digination from the ongoing industrial processes, programs, and personal practices that Nicholas Negroponte described in his 1995 book *Being Digital*. If in Wilhelm's story we get a neutral theory of digital technology that suggests that all of these things are just tools that we put to use in certain ways, with Negroponte we get a healthy dose of hyperbole regarding the emancipatory potential of such technology. I'll challenge both Wilhelm's and Negroponte's stories in what follows, because I think there is much more to these stories than they let on, and it has so much to do with McLuhan's insistence that *the medium is the message*. Next, I want to say a bit more about the title chosen for this book before we unpack these distinctions, and that neat little expression.

Digination (the processes and the thing) also concerns the more immediate psychological and experiential/phenomenological effects, as well as longitudinal social outcomes and consequences of participating in a life-world increasingly rooted in and constituted by digital information retrieval, storage, processing, and exchange. So the process of digitization is really just the beginning, the tip of an iceberg. It is, in fact, only a small part of the wider social processes that constitute our Digination. I'll refer throughout this book to modes of thought and action that emerge from systemic changes in cognitive, experiential, and behavioral patterns happening all around us.

Of course, one of the first and still oft-touted virtues of portable digital communication technology is the unparalleled emergency service access it makes possible. However, conventional talk about the positive benefits of the technology quickly extend to include the conveniences and efficiencies of e-mail, instant messaging, and text messaging systems for the more mundane moments of life. Real-time directions to the best entrance to the shopping mall, a last-minute reminder to pick up some balsamic vinaigrette dressing, a change in plans before meeting friends at the club, and all the rest make up these new exigencies. Ostensive efficiencies and enhancements in the workplace, a geometric increase in the scale of information available to students and practitioners in and outside of academic settings, and a tightening of the familial bond are narratives that still keep many of us interested—and a good fraction waiting with bated breath—for the latest offerings from companies like Microsoft, Intel, AOL, AT&T, Verizon, T-Mobile, Apple, and Google.

The downsides themselves seem always in the process of emergence, but their uncanny ability to go unnoticed by the casual observer continues undaunted. Indeed, the dream of modernity is alive and well here in America, where the "newer = better" mantra is very often read quite literally, and without much concern or skepticism, and so it has many consumers walking lockstep to their favorite high-tech superstore, accessory kiosk, and *sales & service* desk. Several of the negative consequences include a reduction in quality and efficiency at the work place, new hazards in our streets and hallways, crumbling standards of academic integrity, interpersonal conflict, unpredictable and often severely debilitating system failures, new kinds of crime and delinquency, new forms of piracy and theft, widespread privacy and security problems, and a steady disintegration of both the public and private spheres (to name a few). And with all of this the continued blurring and bending of distinctions between home and office, leisure and work, fact and inference, and stranger and friend are only part of an initial list of issues we will have to contend with as participants in this brave new media culture that is our Digination.

So in order to maintain a clear view of what's happening in and around us, a dual sense of the term *culture* will also be required. First is the traditional notion of a social culture where individuals, groups, and institutions interact. Second, it means seeing our natural, built and media environments collectively as a quasi-biological culture where specific media, and media systems find niches, grow, replicate themselves, propagate, and flourish—and may even wither and die—sometimes slowly in a predictable way, and other times quickly and quite unexpectedly. Throughout this *media culture* the role of the individual, while often going unnoticed, will remain crucial. The kinds of media choices we make, as well as the habits we develop with specific media will contribute to and take away from the surrounding culture and wider civilization that is always in the process of becoming. Depending on one's personal tendencies and the context of one's lived experience—that is, where they are both geographically and socially—several examples of the blurrings and bendings just listed might be seen simultaneously as enhancements and constraints, benefit and bane. New methods of observing and measuring brain activity and the burgeoning field of epigenetics (essentially, the study of genetic change that occurs above or after the genome) suggest that the media choices we make today will be consequential with regard to the way future generations will think about, perceive, and interact in, through, and with their worlds.

Another set of realities illustrates how mobile telephony and computer technology greatly reduce the significance of physical place. Increasingly, *where* one is no longer matters. The whole concern about whether spaces and places—be they geographic, social, or conceptual—will retain their delimiting features was a core theme in Joshua Meyrowitz' *No Sense of Place* (1985), and it remains a key question in this book. Now, more than ever, where and how we communicate what kinds of content to whom are concerns ignored at great risk. One thing we wager is an unraveling of the delicate social fabric that is part and parcel to a civil society. My hope is that at least some of the debate surrounding the enhancements and constraints, the strengths and the shortcomings, the improvements and liabilities associated with the regular employment of several popular communication media will be substantially disentangled for the reader by the end of this book.

And so here is a third and final spin we can put on the term. I cheat a little by dropping the second "i" to give us a new word which retains a great deal of relevance for our present purposes. Digination then, finally, alludes to a mounting feeling in many of *dignation* (another noun—French): *the act of thinking worthy, or honorable.* It is predictable that this sense of the term, and that all accompanying sensations, will not resonate as well with those who make up the older, or even middle-aged generations. However, there is a very real sense today especially among the relatively young—that natural part of the emerging technophilic class born in the last 20 to 30 years—who observe and feel that to be without a substantial assemblage of sleek and portable digital devices is to be just a bit less than human. We can align this sentiment, in fact, with Marshall McLuhan's insights concerning the role of the automobile in American culture around the mid-point of the twentieth century: "Although it may be true to say that an American is a creature of four wheels, and to point out that American youth attributes much more importance to arriving at driver's-license age than at voting age, it is also true that the car has become an article of dress without which we feel uncertain, unclad, and incomplete in the urban compound" (McLuhan, 1996; 217).

If young folks these days aren't as in love with their cars as they used to be, they do appear to be obsessed with Wi-Fi-enabled laptops, MP3 music players and iPods, and cool little camera-equipped smart phones. To reiterate, young or old, if a similar set of exigencies is not felt as strongly by some regarding the various physical components of our digital technological apparatus, I have spoken and worked with many people who describe feeling intellectually, even perceptually dysfunctional, and by extension, not quite complete as a person or legitimate as a social agent without the appropriate communication devices in-hand or at least *on board* somewhere: in a pocket or book bag, on a belt, around one's neck, strapped to an arm, or in, on, or around the ear. Perhaps then, from the perspective of the stern social observer, to *be without* is at the very least to open oneself up to social ridicule. At one level, of course, this is one of the intended aims of advertising any product or service. It is the feeling in one's gut that there is something *lacking,* and it is precisely what the marketing establishment hopes the wider population will buy into; the idea that to be without the latest wafer-thin smart phone/camera/ music player or digital media appliance is to be, somehow, in someone's eyes, just a bit less than a whole human being.

At a more substantive and even practical level, however, we find that there may be a mounting truth in all of this. Indeed, what is the status and content of our new ideals? Values are being expressed in the way we consciously think about and ultimately choose the tools we use. Other values are bound up and embedded in the tools themselves. Some of these values are there because the inventors or designers of the tool wanted them to be. Others, we discover, are hidden and quite unintentional. In both cases we are manifesting a certain set of destinies embodied in the different combinations of accoutrements we append to our physical bodies, our minds and memories, and our virtual, symbolic selves. This raises a whole series of pressing questions regarding what it means to be human today. If, for instance, being human continues to include the ability to think, talk, interact with the environment, and engage with others, then we might need to stretch the definition to allow these new tools and mechanisms to be included as necessary, that is to say life-giving, components.

With this possibility, the threat of *autonomous technology* (Winner, 1977) seems to loom large. I make references to technology over-used and out-of-control in several chapters, but in the end I'll argue that it is the media choices each one of us makes that can most reliably predict the deterministic tendencies various media seem to exhibit. We'll see that human intention is sporadically, if temporarily, suspended in the midst of interactions with digital devices and systems of various kinds. Human choice in a field of "intentional machines" is one way we might approach the autonomy question. I'll also introduce a bit more nuance into a common distinction drawn between hard and soft (or strong and weak) technological determinism in an effort to clarify things. A not-too-distant anecdote from the world of business sheds some light on the situation here and the confusion it often generates.

At the 2001 World Economic Forum business summit in Davos, Switzerland, when the meeting theme was "bridging the divides," Chairman and Chief Executive of McDonald's corporation, Jack Greenberg, made a controversial comment. He likened the phenomenon of globalization to the weather. Greenberg said that, as seems to be the case with the weather, globalization "is not something that any of us can change." The media fallout was mixed, with adherents of and detractors to Greenberg's apparently naturalistic declaration taking up sides in neat and tidy rows. Of course, one can certainly begin to protest Greenberg's statement straight away as a strained, self-serving analogy. I could argue, for instance, that globalization is rather unlike the weather since the political and economic policies and structures that enable globalization processes to continue are the products of human decisions and designs, and so they must necessarily be subject to the flights and fanciful dictates of human will. On this view, likening globalization to the weather becomes little more than a gambit that functions rhetorically to strip us of our practical role as planners, decision makers, adopters, and abstainers. If so, Greenberg can be impugned for unscrupulously drumming up support for an unbridled market place that unabashedly aligns the idea of democracy with the doing of unchecked capitalism. But I want to cut to the chase here and suggest that all this hullabaloo relates to communication technology in the following way.

Greenberg's assessment comes close to one interpretation of technological determinism as a universal statement that claims humans build systems they cannot control. As Langdon Winner (1977) suggests in his book *Autonomous Technology: Technics-out-of-Control as a Theme in Political Thought*, most technologies carry with them the tendency to become "tools without handles." Winner goes on to say that, while it might appear that certain people or groups are "at the helm" so to speak, the technology user, even the captain of industry, delude themselves into thinking they are in control when they are, in fact, merely along for a somewhat more comfortable ride.

And yet we could read Greenberg with an even more jaundiced eye. We can look at various natural phenomena like the weather as of late and wonder if human action (sometimes intentional action, but more often not) does seem to play more of a causal role than previously suspected. For instance, while there are still plenty of folks who believe that the activity of human beings in the oceans, on the surface, and in the atmosphere cannot really be detected and therefore could not possibly trigger any alteration in the planet's natural systems, there is mounting evidence which suggests that various forms of contamination, along with temperature change and biomass depletion, are beginning to register on even the most imprecise of scales. Like so many distinctly non-ecological propositions and postulations being advanced these days I'll suggest that such arguments, alongside persistent critiques of technological determinism, are part of a larger rhetorical straw man. What's more, the being-in-or-out-of-control trope really is a false dichotomy. Finding the various junctures and interfaces along a spectrum or continuum of control where individual and collective human activity and action can intervene and potentially change the course of life on Earth is really what this book is all about. Call it co-determinism, co-construction, mutual adaptation, or whatever term is in vogue at the moment, when you look real close, this is just the way the world works. Open systems like our natural ecology and our media ecology are replete with, and constituted by reciprocal formation, relation, and re-formation. I'll argue in what follows that many behaviors characteristic of Digination are heavily dependent upon the way we construct our media ecologies and the communicative niches that emerge out of them.[1]

With digital media coming into their own we are each, increasingly, standing at the center of a communication network and information landscape. However, if we are both nexus and node from a functional standpoint, a media ecological perspective does not place us at the center of a communication environment without also giving us the potential to alter the course of things both locally and abroad. When taken to heart, an ecological view of media environments provides us with new powers of observation, and a new sense of place in and relation to our social, symbolic, and physical environments—both the natural and the human made. A media ecological perspective informs our ability to be literate technology users. And borrowing a term from cognitive psychology, I'll suggest why a new media literacy must include developing an awareness of something J.J. Gibson (1979) dubbed *affordances* (practical, strategic opportunities for action) that exist in all environments—whatever their organization and make up.

Being media literate in an ecological sense includes giving serious consideration to, and taking the appropriate advantages of, any number of these "natural" affordances in one's environment. For example, paralleling Aristotle's definition of rhetoric as *the available means of persuasion* with a rhetorical perspective allowing one to see the world in a strategic way, we can see how (if I'm oriented to my environment in the most beneficial way), my coming upon a tree stump in the middle of the woods affords me the possibility of a table, or a seat, or maybe even firewood if I have an axe or hatchet at hand. In the same way Aristotle conceived of the many things and concepts in a speaker's environment that can be used to create persuasion, Gibson's notion of affordances assumes an innate kind of awareness. Wild animals live by this awareness, but most humans, with their built environments, have to learn their own theories of affordances with each new niche they encounter. It is the awareness that using certain *things* will often allow you to achieve your goal or make you more effective at a given task, than using certain other things. So, considering that tree stump again, I can recall the advice a very musical colleague of mine once offered when I had designs on building my own guitar. She knew, for instance, that Brazilian Rosewood would make a better sounding and feeling guitar than Pine.

As the communication scholar Richard Weaver (1953) pointed out, "[r]hetoric has a relationship to the world which logic does not have and which forces the rhetorician to keep his eye upon reality as well as upon the character and situation of his audience" (p.53). A media ecological perspective is like a rhetorical perspective in this way. The audience, as actors taking up a functional role in the unfolding story with and through the media they choose, make all the difference in the world.

Given this, the definition of a media-literate or media-savvy person should shift from describing the ability to ostensibly manipulate a technology or some combination of technologies, to describing someone who is able to recognize when and how a given technology or set of technologies aids and/or inhibits the accomplishment of some specific task at hand along with what sorts of wider effects or side-effects might be associated with the use of that thing. Beyond this, the *new media literate* will have learned to see how the tool often fundamentally changes the user, the task, and the relationships between all three. In other words, the best new media users will have to engage in lots of picking and choosing in order to accomplish things in the manner desired. They'll need to not only be aware of what certain media can do, but undo as well. Toward that end, it might be worthwhile to start talking about sustainable communication practices and media use in ways similar to how we now talk about the sustainable use of energy and perhaps sustainable consumption in general.

Two very mundane examples can illustrate the problem well enough. Consider the default behavior in business offices and classrooms over the last fifteen years to use *Powerpoint®* for a presentation because that application happens to be installed on the computer on hand. This sort of standard operating procedure might need to be rethought. Not to pick on Microsoft too much here, but we'll also have to reconsider the showing of business plans, quarterly projections, and class presentations in *Excel.* Why? Because in simply doing so, in just using Excel because that is what's available, and then moving on to the next item during a meeting since it's fast and efficient to do so . . . and because, well, we assume that our audience now "has it," is nothing less than narrow and faulty thinking. Or, as McLuhan (1964) liked to say, it is "the numb stance of the technological idiot" (p.18). One problem with this kind of activity, as well as the assumptions that accompany them, is that a message sent does not necessarily equate to the same or intended message having been received. Aside from digital transmission (and some would even argue this) there is no such thing as isomorphic communication—or the perfect transfer of meaning. In the same way urban planners, transportation specialists, and logistics experts grapple with something called "the last mile problem" when considering augmentations to multi-modal urban transit systems, we might begin thinking about a sort of "last mile problem in communication" as a problem of multi-media selection. Media don't just send us, they alter us along the way.

This new kind of literacy can help us see how certain media and technologies can be good for certain things, though not all. In filling out his "Huxleyan warning," Neil Postman (1985, 1992) concluded that television was good for one thing and one thing only: entertainment. Extending McLuhan's insights on the subject, Postman even went so far as to say that given the visual bias of television and the subsequent proneness toward emotional processing the medium entails, any time we try to convince ourselves that television can be used to relay serious, rational thoughts and ideas we inevitably play the fool. If Postman's assessment does sometimes seem a bit extreme, he was surely on to something with that sentiment. Today the majority of Americans employ and happily watch television to do and find out about so many things. We certainly amuse ourselves with light fare including all manner of cartoon and camp now beamed via satellite and releyed via co-axial and fiber-optic cable 24/7 on the comedy channel and cartoon network. In similar fashion we watch and post e-comments on game shows and talent search programs. We share inside jokes with Jay Leno, David Letterman, John Stewart and their ilk. Watching sports is great fun too. No problem there thought Postman. These are all the things TV does best. The problem for Postman is that we also try to seek out news about an increasingly complicated world, or learn about our political candidates, or "attend" church or participate in a memorial service through our televisions too.

So then the *globalization-like-weather* proposition actually retains some utility as an analogy for a *Digination-like-weather* idea. To be sure, unless we take the appropriate steps in the construction and maintenance of our digital *clothing, architecture,* and *environments,* we risk being left subject to the intentions embedded and encoded into these powerful ideas and objects. Indeed, the weather, globalization, Digination . . . they all seem, in their own ways, to be forces of nature.

I think the analogy probably works even better when we talk about the process of Digination being like the weather inasmuch as it's going to happen whether we like it or not—the question is *how.* If the younger generations seem to be embracing their new environments wholesale and without much question, at the fringes we can still find people who appear to live a relatively unmediated existence. Even if we consider Postman's warning and give the process more than just our mindless inattention, we are still participating, and Digination is still surrounding us and so happening to us all the time. In such a situation our physical representations are just about the only aspect of our identities that we retain any real control over because our digital and otherwise disembodied symbolic representations are subject to the digital data processing infrastructure that performs near constant quantitative manipulations of those representations. In fact, they are part and parcel to that infrastructure. And so this brings us to the unique contribution media ecology and, by extension, biology can make to the present study—and media studies in general.

MEDIA ECOLOGY

As practitioners in the "master discipline" of biology—the *science of life*— biologists are interested in where organisms originate, how they maintain their structure and function, how they grow and propagate, and how they have evolved and continue to evolve over time, or die off. Logan (2007) has called for an explicit biological approach to help systemize the broad area of theory and research that is *media ecology*, an interdisciplinary field that typically combines elements of biology, philosophy, psychology, cognitive science, computer science, anthropology, sociology, history, and economics, along with cultural studies, technology studies, and media studies. Media ecology has been dubbed a "meta-discipline" and still remains a "preparadigmatic science" (Nystrom, 1973). As a meta-discipline, media ecology is an expansive, integrative, *over-seeing* approach that seeks to find the many sensible but often subtle connections between ostensibly disparate modes of analyses concerning equally disparate subject matter. This is done in an effort to draw out the contours and consequences of an almost infinite combination of human-media interfaces and fusions with technology. However, as preparadigmatic scientists, media ecologists still do not, "as yet, have a coherent framework in which to organize their subject matter or their questions" (ibid).

This book is part of that collective project to which Nystrom refers: an effort to help organize things. While I will not cull on specific biological principles in any systematic way throughout the various chapters, I will periodically point to the manner in which much of our intermingling with communication technologies of various kinds has a morphological and therefore evolutionary component to it. That is to say, we are changing something about ourselves as we continue to invent, integrate, and inhabit digital artifacts, processes, systems, and procedures. Consider, as a superficial (physical) example, the way thumbs have become the primary digit used to manipulate the standard phone keypad game controller, and miniature keyboard for so many in our younger generations. More significantly, perhaps, we should consider how the cell phone's memory has taken over most of the function of the biological memory with regard to recalling even the most frequently called numbers. Since biology includes the study of life functions and processes within and between organisms interacting in their environments, a biological approach might be a particularly good way to understand these kinds of phenomena.

Again, while periodic reference to biological terms and concepts will be made throughout this book, the biological is not being marshaled in order to make any essentialist or naturalistic claims about the status of certain groups of people or technologies. Nor will I highlight any particular group of people using certain technologies as the ones to follow. Instead, drawing on survey, ethnographic, archival, and participant-research data, I proceed in each chapter by making a series of observations on the particularities and idiosyncrasies of certain media-in-use that do at times appear to mimic biological patterns and processes. I admit to engaging in some inference and conjecture based upon the data, however I do so in an attempt to draw out the various practical benefits and constraints bound up in the technologies so many of us find ourselves not just employing, but in a very real sense *integrating with* every day.

As detailed in the next chapter, a well-established foundation already exists for this way of thinking about technology. Marshall McLuhan popularized the systemic approach to understanding media and their interplay with culture and technology. However Harold Innis and McLuhan's other key intellectual mentor Lewis Mumford, along with a collection of thinkers working in as many disciplines, including Norbert Wiener (1948, 1950) and Jacques Ellul (1964) all proffered system-theoretic views regarding the human use of technology and, through reciprocal relation, the use of human beings by technology. In an excellent essay about Lewis Mumford's *Technics and Civilization* (1934) Andrew Jamison describes Mumford's system-theoretic approach to understanding technology. I like the following passage, where Jamison does a nice job characterizing the ecological and ultimately biological framing of Mumford's work.

> Mumford was unique among American intellectuals for combining what were already separate fields of inquiry, distinct specializations—science and technology on the one hand, culture and society on the other. The two cultures did not exist for Mumford in separate spheres; as a boy he had enjoyed fiddling with radios as much as reading classical literature, and for most of his long life, he saw his main task as bridging the infamous two cultures or at least bringing inquiry about them—together. He didn't combine the cultures by reducing one to the other, but by transcending them both and operating on what might be termed a meta-level of reality, where totality exists. It was by trying to be all-encompassing, by seeing the world in terms of patterns, processes, cycles, that is, by adopting an organismic world view that one could overcome

specialization. In this respect, Mumford was inspired by Whitehead, as well as by Patrick Geddes, in thinking of society and its activities through biological concepts, in terms of life processes.

(www.easst.net/review/march1995/jamison.shtml)

I think this passage does just as well in framing my own inquiries. Functionally speaking, digital communication technologies enable a kind of near-instantaneous disembodied communication and transportation, with fax machines, cell phones, e-mail and an Internet presence enabling people to get things accomplished at a distance. These apparatuses can transport our voices, our thoughts, our signatures, and even our physical likeness when we cannot (or do not wish) to be there in person. At yet another level, then, Digination is about the real-world transformations that are accompanying such virtual transportations.

Of course any effort to explain these processes is going to run into the problem of under-determination. We will never know all the connections and causes. And there are, clearly, some limitations to the analyses undertaken here. Nonetheless, thinking in ecological-qua-biological terms about the many communication technologies we regularly employ is a potent approach that has yet to be systematically applied to real-world media use. A media ecological perspective that explicitly employs biological concepts illuminates processes of Digination to show how these tools function in altering the various ways people and societies come to *see* and *be* in their worlds. In short, a media ecological approach to understanding our Digination allows us to assess the fitness and predict the sustainability of digitally mediated relations, as well as the health and sustainability of the myriad social, educational and political entities and institutions progressively finding themselves constituted by, if not actively constituting themselves through, such relations. An intentional media ecological outlook can offer valuable guidance concerning what it is about our communication tools and our relationships to those tools that we might want to change (as well as what we might not). The decisions we make in this regard are consequential to everything from participating substantively in anonymous meet-ups, to building quality professional collaborations, to sustaining fulfilling and worthwhile personal relationships, familial interactions, political parties, and national identities.

Finally, understanding our Digination along these lines will provide valuable insight into what is currently happening in our media environments, and how that might have something to do with what is happening in and to our natural environment. I address this concern briefly in the penultimate chapter, which is devoted to an analysis of *eBay* as a socio-technical system. I then revisit this relationship between our media ecologies and the natural ecology in the final chapter, where the biological approach is described and illustrated in much more detail.

PLAN OF THE BOOK

Having unpacked the concept of Digination, provided some personal history concerning my participation in it, and offered a preliminary sketch of the meta-theory framing my inquiries, I want to next outline the basic progression of the book. Chapter 2, *Lost Logos: Finding the Art and Argument in McLuhan's Message,* offers a little background, and some insight into the man who really came up with this idea, or at least foresaw the effects of Digination. I'll discuss some of the intellectual history related to Marshall McLuhan's rise, fall, and recent re-emergence. I then detail several of the ways I will marshal McLuhan throughout. Aware that I'm always on the verge of spreading him too thin, I approach the various media under analysis in these chapters with a broad interdisciplinary eye. In the end, however, I think the reader will agree that this allows for a pragmatic understanding of the communication technology that surrounds, carries, and increasingly constitutes us today.

Chapter 3, *Indigenous E-mail: Identity Construction at the Oral/Textual Interface,* follows the experience of a unique group of digital immigrants as they try to make sense of what is for so many digital natives today a mundane (if obsolescing) communication technology. First, the structural features of a basic e-mail system are described to demonstrate how the neutral theory of technology—the standard trope which suggests that it's how we use tools that matters—is highly suspect.

The second half of the chapter contains a case study detailing the experience of several members of a Mohawk (Iroquois) Indian community who began using e-mail for the first time in their collective history in the late 1990s. Most of these people soon reported tensions between the way they wanted to communicate and the way their communication "came out," "sounded," or "felt" while using the e-mail system.

A number of features of the e-mail system, including the subject field, a few user prompts associated with the subject field, the composition window, and the linear, sequential format of the medium, are identified by users as problematic. While features of the medium seemed to play a role in this scenario, it is the observations made by members of the group about themselves as *Mohawk users* of e-mail specifically (and computers and the Internet more generally) that became significant symbols in their sense-making process. In this sense the chapter details an ethno-methodological analysis that stays close to Harold Garfinkel's original formulation. I think the chapter also works well as an introduction to some of the "standard" processes of Digination as it describes the relationship between user and machine, and self and other that emerged at the digital interface.

In Chapter 4, *Blogs: The New News Medium*, I extend this discussion of digital interfaces by focusing on the psychodynamic effects of textuality and disembodiment as they function in the consumption and use of web logs dedicated to the dissemination of news. News Blogs have become the default political news source for a growing number of well-educated and ostensibly well-informed segments of the population, and bloggers and blog advocates suggest that these digital manifestations offer something different and potentially unique to the twenty-first century citizen. At their best, blogs represent a new form of open-sourced/open access partisan press that promises to bring McLuhan's tribal context—his *global village*—one step closer to fulfillment: a vibrant, interactive polity resembling Jurgen Habermas' "Ideal Speech Situation" where differing minds engage one another without fear of censure or reprisal.

However blogs, due more to certain structural features than any specific kind of content, might also represent the latest form of mass-mediated triviality and celebrity spectacle, with the potential to create and sustain insulated enclaves of intolerance predicated on little more than personal illusion, rumor, and politically-motivated innuendo. With a more social scientific bent than the previous chapter I unravel some of the personal, social, and political significances of blogs and blogging as a pre-eminent manifestation of Digination.

In Chapter 5, *Information, Interactivity, and the Denizen of Digination,* I broaden the discussion to consider the epistemological or meaning-making function of online news and information platforms that are moving progressively from digital text to imagery. As in the previous chapter, I argue that certain properties of digital encoding, processing and retrieval are prompting a shift in the form of online content from words to pictures as the operative informational unit. This shift is described as a natural side effect of Digination and I suggest why there are some pressing issues bound up in the fact that so many more people are born today into an environment heavily biased toward the image and other forms of pictographic information.

In the early 1990s, the proliferation of the Internet, with all of its personalized information services, photo news galleries, computer simulations, and interactive media links on educational, governmental, and commercial Internet news and information sites, was highlighted by President Bill Clinton and, more infamously perhaps, by Vice President Al Gore as one remedy for this troubling state of affairs. Presidents Bush and Obama continued the argument for a nation enabled by digital communication technologies. I'll argue, despite political, professional, and popular opinion to the contrary, that as the Internet becomes more technically sophisticated, a proportionate, though inverse trend in the epistemological sophistication of its user base may be inevitable. I then discuss some of the potential implications this trend holds for the future of a global citizenry.

Chapter 6, *Search Engineering and our Emerging Information Ecology* extends the discussion in chapter 5 regarding technological form, capacity, and usage patterns. I focus the inquiry on artificial intelligence constructs designed to aid human users in their information gathering and assessment tasks. *Interface Agents, Infobots,* and *Knowbots* are terms that have been used to refer to artificial intelligence systems designed to perform automatic searching functions that also parse, process, and compose natural language streams without human editorial input. In terms of building a collective epistemology, these constructs are shown to exert more agency and have more causal force than we might suspect. They are some of the new denizens of Digination.

The chapter begins with a comparison of two of the leading web browser platforms: Google and Yahoo. I consider how the proprietary algorithms underlying these reigning platforms produce information, muddy the distinction between source and channel, and help define what we mean by an "authority" online. Finally, as in the previous chapters on Blogs and Internet News, I question the social and political utility of information gleaned from these information constructs as they continue to fill out the contours of our Digination and begin to constitute a primary habit of thought in its inhabitants.

Chapter 7, *The Sound-Tracked Lifeworld,* investigates the nature of our digitally enabled soundscapes. The radio was the first technology to allow the individual to become enveloped by the familiar and exclude the annoying, alien, or otherwise unwanted auditory experience. In all its guises, the personal digital music device (PDMD), with its combination of extended battery life, massive memory capacity, and random play function is something wholly new which may require a reformulation of our traditional psychological understandings of space and time. By turning on their iPods and other PDMDs users are now able to, in effect, turn off all that surrounds them and engage in programmed or impromptu theming of both novel experience and daily routine.

Aided by several research participants, I explore the various ways in which the PDMD alters an individual's phenomenal relationship to the world and in so doing simultaneously prompts a conceptual reconfiguration of physical space and social place that also typifies our experience in Digination. With all of the wonderful things our digital soundscapes afford us, I caution against the kind of indiscriminate consumption so many folks engage in and, more crucially perhaps, the incessant consumption of music and sound that many engage in when occupying public spaces and other social contexts. We'll see how these devices may be contributing to a situation where a new breed of social minimizer is emerging, who walks and runs ghost-like through the world.

In Chapter 8, *Podcast Consumption: From Sound Track to Narrative Track,* I continue the discussion of mobile listening. Instead of the mobile soundtrack, however this chapter focuses on the unique powers of the human voice and the capacity of the mobile narrative track to direct thought and experience. Extending the discussion of the social effects of blogs, aural digitality, and info-mobility, this chapter details the growing popularity of podcasting and the manner by which personal perspectives and preferences blur the nature of the public and the private realms to unprecedented degrees. In ways that seem akin to the work of a DJ writ large, a relatively new concept dubbed *publicy* perhaps exemplifies this phenomenon. Publicy (Federman, 2003) describes privacy occurring under the intense acceleration of instantaneous communication. I argue that the podcast is one of the purest forms of publicy then wonder if this really is something new or just a new form of orality (i.e., *secondary orality*) that undergirds our emerging Digination. Podcasts recast the listener as a captive and willing witness to the chants and decrees of the twenty-first-century-village elder, the village idiot, the iconoclast and cultural mystic, and the vitriolic ideologue.

In Chapter 9, *Knitting, Napping, and Notebook Computers (and other mnemotechnical mechanisms)* I begin by pointing out that Wi-Fi technology continues to be installed throughout our classrooms, public libraries, coffee shops, and state parks at a blistering pace. I then suggest why a number of intriguing questions remain unanswered. For example, how efficient and effective is this culture-wide project? Extending the general critique in chapters 5 and 6, are we enabling our population as hoped? The contexts of use are so varied, however, I focus most of my inquiry on the classroom situation. I also consider briefly what others tend to be doing with their computers, PDAs, and smartphones at the dinner table, in their cars, at the office, during meetings, and in every other corner of our Digination. I'll again close the chapter considering some of the likely cognitive and cultural consequences bound up in our interminglings with these digital appendages.

In Chapter 10, *eBay Ethics: Prefiguring the "Digital Democracy,"* we take a tour of what is arguably the fastest growing intentional environment (and economy) ever conceived. I begin with a description of the core features of *eBay's* user interface. With the aid of several co-researchers I illustrate how in digital contexts like *eBay* there is a tendency and even a built-in bias toward top-down (i.e., corporate, quasi-governmental, administrative) entities. While the functional role and intentions of these entities quickly and reliably diffuse into the user base, they nonetheless continue to effectively enhance their ability to rationalize and control an otherwise seemingly democratic, bottom-up, and peer-to-peer situation toward their own narrow interests. The analysis highlights important relationships between the concepts of citizen, consumer, and political actor today, and speculates into the significance these various social roles might play in a full-fledged *digital democracy* of tomorrow. I suggest that *eBay* might even have all of the requisite features aspired toward in one possible instantiation of a total and complete Digination, and end the chapter speculating on some of the ways activity in virtual domains like *eBay* may have deleterious effects elsewhere both on- and off-line.

Chapter 11, *Media Ecology and a Biological Approach to Understanding Our Digination,* is the final chapter. I begin by reminding the reader of the three themes that comprise the subtitle of this book: *identity, organization,* and *public life*, and briefly review where and how these themes were threaded throughout. I suggest that we remain cognizant of the changing nature of identity, the different forms of social, cognitive, and political organization now emerging, and what it means to be a *world citizen* today. I then wave my hands in the direction of a new methodology for media researchers, and a new kind of ethnomethodology (or way of living and making sense of reality) for all of us living in the digital age.

I close the book by filling out the discussion initiated earlier in the present chapter regarding media ecology. Again, the procedural map for this project, and the primary lens through which I view the way people relate to media (the media ecological approach) was pioneered and promulgated by Marshall McLuhan, Edmund Carpenter, Susan Langer, Elizabeth Eisenstein, and others. I draw out the environmental significance of this approach, and make explicit many of the natural affinities between biology and media ecology.

Throughout the book, I allude to this different way of thinking, one intimately related to a media ecological outlook regarding the human use of technology, and technology's use of human beings. A biological metaphor (and perhaps more than just a metaphor) can show us where we've been, illuminate the current state of our affair with digital technology, and even offer some fresh perspective concerning where to go next. Two hundred years from now, what will someone think of this moment in the history of the world? Will it be seen as a fairly unremarkable stage in the progression toward some ultimate, ideal intertwining of human and machine? Or, will it appear as just a short-lived digression, a time when we lost track for a spell of our sense of connectedness with each other, with ourselves, and with the Earth under our feet? In an effort to engage and channel that tenacious striving toward balance and sustainability that is the hallmark of all biological systems, I suggest in the chapters that follow how media ecology can offer some sensible and productive ways to proceed from here on. Indeed, this may be our only hope in building a future worth living.

NOTES

1. Developmental Systems Theory (DST) is a new area of inquiry in evolutionary biology and ecology and seems to center upon epigenetic processes, as well as processes of *niche construction* that make up our natural ecology. For some fascinating and fruitful twists on our traditional understandings of evolution that promise to become core methodological components of media ecology see Oyama, Griffiths and Gray (2001), Odling-Smee, LaLand and Feldman (2003), and LaLand and Brown (2006).

Chapter 2

Lost Logos

Finding the Art and Argument in McLuhan's Message

"We're the first culture in the history of the world that ever regarded innovation as a friendly act."

—Marshall McLuhan

There's a pithy cartoon in an old dog-eared communication theory textbook I've had on my shelf for years. It depicts a somewhat older man with thinning hair huddled over a broad desk strewn with books and papers. He has a worried look on his face as he ponders these materials in a well-stocked, if somewhat cramped, private study. A classically dressed woman of similar age stands in the doorway with a blank expression on her face. The caption under the cartoon is her imploring request: *"For heaven's sake, can't you just relax and enjoy art, literature, poetry, history, and philosophy, without trying to tie it all together?"* While I have yet to determine if the cartoonist was directly referencing McLuhan and his wife Corinne in the image, for me this still captures McLuhan's method, and maybe his role in history too. In the next several pages I want to say a bit about McLuhan's method, his place in the history of ideas, and how he can function as a kind of tour guide for anyone interested in (or already certain they have a knack for) thinking about communication technology today. For novices and experts alike, I think McLuhan has some valuable things to say, and I do my best throughout this book to clarify his ideas and insights.

The publication of *Understanding Media* in 1964 quickly garnered the iconoclastic Canadian English professor a broad lay following as a well-read student of history, a keen observer of the present, and a prescient seer to possible futures. Many academics found McLuhan's thinking attractive due in part to his interdisciplinary approach: weaving literature, history,

25

philosophy, technology studies, psychology, and the still nascent field of cognitive science—among other things—into his analyses of communication media.

The emergence of sporadic pockets of cross-fertilization and disciplinary integration in the academy in the late 1960's seems to have been a key factor behind McLuhan's popularity at the time. However, while popular interest remained fairly constant over the next several years, McLuhan, and his many *McLuhanisms* (like his ubiquitous "The Medium is the Message" and "Global Village" ideas) steadily lost traction in the United States in the early 1970s. A subsequent push back toward disciplinary compartmentalization and increased specialization in the late '70s probably has much to do with a series of gross misreadings of his ideas at the time. A concomitant shift in the popular mind that read "high-tech" as a new mantra and looked specifically to new and emerging computer technologies as the progenitor of positive cultural changes no doubt added to the skepticism surrounding McLuhan's thoughts on the matter. However, the cycle continued and a return to interdisciplinary studies in the waning decades of the twentieth century reinvigorated interest in his ideas and generated new and largely coherent interpretations of McLuhan's mercurial thinking.

McLuhan popularized the idea that technologies ranging from roads, numbers, clothing, money, clocks, the bicycle, automobile and airplane, and the telephone and television (among others) could be thought of as extensions or prostheses of the human beings who use them. Humans have always maintained close relationships with tools and mechanisms of various sorts. And today many people who are regular and adept users of digital communication technology will often admit to feeling as though their machine is an essential part of them or, perhaps, they feel like some small part of a machine. There is something to these feelings, as there is something *intrinsic* about the devices we live and work with.

At bottom, McLuhan's media ecology is about finding in the midst of all these potential prostheses, sustainable relationships and ratios between the six exteroceptive senses (seeing, hearing, taste, touch, smell, and balance) along with proprioception (the sense of relation between various parts of the body) and even interoception (the internal sense of pain and movement of the organs). Traditionally, the sustainable relationship has been understood to mean that which is capable of continuing with minimal long-term [negative] effects on the environment. Concerning communication technologies, this might include informational, psychological, cognitive, social, and political environments.

Despite some marked progress over the last couple of decades in media studies, human-computer interaction research, psychology, and cognitive science toward understanding the role communication media play in our lives, there are some concerns that a kind of disciplinary contraction in the broad field of communication is again in the offing. We find many departments and directors calling for more boundaries and concentrations in the quest for legitimacy, funding, and students (who today have one eye on the syllabus and the other closely scanning the job market). But this is intellectually counterproductive and may be problematic in much more profound ways since communication is evolving to encompass newer subject-domains. What's more, there seem to be some real, practical connections possible between the various fields. James Carey points out another "advance McLuhan pioneered, and which set certain constraints upon his critics grew directly out of his literary studies. Students of the arts are likely to examine communication with quite a different bias than that advanced by social scientists" (Carey, 1998).

McLuhan's writings have always felt more like art than science to me, and I think this was the experience of many students and scholars who were pointed in his direction by astute mentors, or just discovered him on their own and tried to grapple with his ideas as best they could. A slow renaissance in *McLuhan studies* began after his death in 1980 and continues today. It is certainly a treacherous renaissance to be involved in, both practically and professionally. Long clauses sprinkled with aphorisms and puns, all interspersed with what are often over-extended examples and analogies, his writings can seem more aphoristic than argumentative, more polemical than pragmatic. McLuhan's method was intentionally interdisciplinary as he was inspired by such associative and integrative thinkers as Patrick Geddes, Lewis Mumford, Harold Innis, and Teilhard de Chardin. Of course, the genesis of such an approach goes all the way back to the ancient Greeks—and McLuhan knew them too.

Popularized by McLuhan in the late 1960's, the media-theoretic approach actually has very deep roots. The idea that a communication medium tends to causally override the content it carries and exert significant effects on the human sensorium was first articulated in a formal way nearly 2500 years ago by Plato in his dialogue *The Phaedrus*. In that document, Plato—extending some of Socrates' own opinions on the subject—suggests that the steady proliferation of the written word (then manuscript texts of various sorts) would be accompanied by a progressive bias toward the eyes to acquire information about and from one's world. And the eye would do all of this at the necessary expense of other senses. What's more, due to the potential writing has to persist in the visual field; people would inevitably come to rely on the medium as a kind of extension of the mind. The result would be, according to one character in the dialogue, a false sense of knowledge, memory, and wisdom.

Due in part to an over-abundance of literal interpretations regarding the Platonic dialogue, media theory is easily misinterpreted or oversimplified to imply a strict form of technological determinism, when it really seems to be about inoculating against the common tendency to think that technology is neutral: *it's not the tool, it's how one uses the tool that matters.*

Instead, McLuhan and others of that ilk seem most interested in getting the idea across that all technologies have causal efficacy. That is to say, technologies can prompt actual changes over time in the physical make up, as well as the perceptual and cognitive processes of technology users. If this still sounds overly deterministic I'll suggest why and how McLuhan was not the strict technological determinist he's made out to be in so many popular critiques.

After a protracted academic upbringing McLuhan enjoyed a fairly brief time in the spotlight, with the 1969 Playboy magazine interview perhaps signaling his apex in the popular press. Over the next decade McLuhan was, at best, trivialized by lay audiences, glossed by literary critics, and rebuked as a charlatan by many of his academic peers. I try to channel the best of McLuhan in this book. But I have to admit I only *think* I know what I'm doing here. In the chapters that follow I make some educated guesses toward emulating what McLuhan's method might have been had he lived to see such things as blogs, iPods, Google, and eBay. Consider the way McLuhan describes his "mosaic method" in the *Gutenberg Galaxy*, the first of several definitive works.

> [T]he galaxy or constellation of events upon which the present study concentrates is itself a mosaic of perpetually interacting forms that have undergone kaleidoscopic transformation—particularly in our own time (McLuhan, 1962, p. 7).

The vertigo I still sense when reading this passage, and the dizzying state of affairs McLuhan alluded to when writing it has not subsided. There has been no calm in the storm of our changing media forms. Nor has there been any respite in the diffusion of innovations that simultaneously prompt and promise to contend with those changes. There has also been no appreciable lull in our attempts to make sense of it from so many discrete perspectives, including the economic and anthropological, the philosophical and psychological, and the scientific and religious in the nearly fifty years since.

McLuhan's methodology in *The Galaxy* and elsewhere was against the prevailing methods of the time. It was much more a kind of counter-method. And there was, I think, good reason for it—a couple in fact. The first has to do with the reality of the world he lived and worked in. Throughout the 1950s and well into the 1960s, the academy was typified by a fairly rigid collection of disciplinary compartmentalizations. Academic specialization and the cordoning-off or "siloing" of secondary and higher education is, arguably, one unfortunate side effect of print technology with its linear, sequential, cause-and-effect structure. I don't think it takes a *McLuhanite* to see that academic specialists are one of the natural by-products of two key features inherent to typography: lists and categorization. But there were other forces in play.

After World War II and into the cold war academic research shifted to high gear. If initially the product of literate culture, academic specialization was also seen as the intellectual's best recipe for success because the technological race was on—a race with some fairly specific goals and even a few measurable outcomes. Technological determinism became what I call a "live theory," that is, both a description of a certain state of affairs and a prescription for action. The finish lines were being laid down: tighter control of the atom, unraveling the secrets of DNA, getting a man into space, and on to the moon, etc. Though something was lost in the nuclear race, the genetic race, and the space race. McLuhan understood what was lost, put off kilter. He also had a pretty good idea of what was needed to restore balance to technological and cultural production that always run hand in hand. What we needed most was to get to, or perhaps get *back* to, a more holistic, more systemic understanding of technology's intimate relationship to culture—and vice-versa.

And that's where books like *The Gutenberg Galaxy, Understanding Media, The Medium is the Massage,* and *War and Peace in the Global Village* were supposed to fit in. There is no question that McLuhan was also a bit ahead of his time. Or, in any case, he was certainly thinking and writing in an age before we had the kinds of technological innovations that might have helped him get his message across. McLuhan was thinking beyond and outside of the book and, specifically, the written words that make up the book. However books were the only medium academics really had at their disposal to make their case and show their wares. Unfortunately, the literary form ended up hoodwinking many of McLuhan's readers into reading him literally.

I happen to think some of that literal-mindedness, and some associated bewilderments that are part and parcel to that mode of inquiry, was also the consequence of lazy reading and research. Then again, Jonathan Miller's (1971) confusion of the notorious "hot and cool" distinction points to a feature of McLuhan's work that remains, for the most part, opaque to so many who struggle to understand his allusions, examples, and illustrations. I think the remainder of the confusion stems from a misunderstanding not of McLuhan's words, but rather, his aim. And, again, books were probably not the best media form to carry and constitute his words or aims. He may have sometimes been frustrated with writing to the point where he was intentionally vague in an effort to perform a basic heuristic function for his readers.

McLuhan seemed to want, above all perhaps, for his work to be accessible to the first post-literate generation. This is why he started to use pictures and other graphical images that took up much more space than his words, with many printed askew, upside down and/or backwards in pocket-sized books. He used his mosaic method (his readiness to borrow insights, examples, and conclusions from any discipline or genre, from anyone, at any time) and deployed his probes (also called McLuhanisms: those bumper sticker quips and alliterated aphorisms) in the hope of getting his ideas across to, into, and back out of others in new and different forms. I'd say McLuhan was interested in progress of a very particular kind.

Indeed, when it comes to technology, McLuhan argues that real progress manifests not just in the production of and ability to use certain things, but in the understanding of those things we produce. I like to think that McLuhan's method was always about a journey—a journey of understanding. And it was always an adventure too—one without a clear and explicit map, the very best kind in my opinion. And so this book is also a journey of understanding. It began as a personal, daily journey I travelled as a graduate student in a hybrid doctoral program in communication and philosophy in the mid-90s. It then became a set of annual pilgrimages to the national and regional communication conferences. It is also a journey I travel throughout the years with my students. It's a journey that brings us to new understandings of the relationships between the things we produce and use, and the kinds of beings we become in the process—a process of mutual production. Or, as John Culkin, a friend and colleague of McLuhan's liked to say: "We shape our tools and thereafter they shape us" (in Stearn 1968, p.60). So with all of the things being shaped these days, our efforts to understand what the hell it could all mean must be guided by the only principle that does not inhibit the progress of understanding, and that is: *Anything Goes* (Feyerabend, 1975). Psychological, economic, and computational approaches to understanding media have had their turn. An artistic-philosophical-literary-socio-historical-quasi-biological approach like McLuhan's should be given a whirl.

Philosophers like Paul Feyerabend, Alfred North Whitehead, and before them Henri Bergson, Johann Gottlieb Fichte, David Hume, and Heraclitus all said much the same regarding the nature of reality. For all of these thinkers the world was, in a word, very messy. Biology, the logical underpinning of media ecology and the modern articulation of McLuhan's method, has always been very messy too—with emergent properties that are hard to grasp.

To be sure, the world is not a billiard table. We are now deep into an environment of non-linear relations, where the law of cause and effect remains, but only to be perturbed by plurality, clouded in equifinality, aggravated in multiply—realizable features and forms, and punctuated by some very mysterious kinds of emergence. With some adjustments McLuhan's "tetrads" (McLuhan and McLuhan, 1988) or laws of media approach to understanding media is a particularly useful way to ferret out many of these non-linear relations and emergent phenomena. The tetrad simultaneously illustrates some of the social-psychological, cognitive, and communicative roles a technology plays—and plays out—in a culture.

> Can the "laws of media" described in Marshall McLuhan's posthumously published writings (McLuhan and McLuhan 1988, McLuhan and Powers 1989) help us understand the effects of the Internet and the new real-time technologies on culture and society? Perhaps the overambitiously phrased and deterministic laws of media can contribute to the creation of distinctions between different kinds of effects; if so, it can be worthwhile to try them out. So far in the public debate on the Internet, contributors tend to argue, simplistically, either that the Internet and related technologies are necessary conditions for true democracy or that they entail the end of democracy; that they obliterate hierarchies or strengthen them; and that they either homogenize or heterogenize people— render them more similar or more different. All points of view may be true, but they apply to different social fields, and at the end of the day, they are wrongly phrased. McLuhan's model can be a starting point (although it needs supplementing) for a more accurate reflection on these issues. (Eriksen, 1996).

Eriksen highlights a conceptual weakness in McLuhan's general approach and makes an excellent point regarding the practical utility of the tetrad itself; it's that McLuhan's project needs focusing. Specifically, as Eriksen points out, the tetrads need to be applied to "different social fields." This is because the effects of various media are themselves unique (though sometimes redundant) and impact differentially across the technological, social, cognitive, cultural, and even political realms.

McLuhan's tetrads or "laws of media" are heuristic devices that end up being very useful in analyzing, and ultimately graphically representing the four laws (or law-like effects) of any medium or technology. According to the tetrad, all media: ENHANCE (extend or intensify) something; REVERSE (or transcend themselves), by "flipping into" something new; RETRIEVE (or bring something back from the past); and OBSOLESCE (force some things to go away). But the generic tetrads McLuhan (1977) originally postulated were only a kind of first-pass, global analysis of the effects of a given technology. This is still largely the case in McLuhan and McLuhan's *Laws of Media* (1988), where almost the entire second half of the 250-page volume is devoted to "tetradic analyses" of everything from *alcohol, high-rise buildings*, the *refrigerator*, and *sewer pipe . . .* to the *cliché*, the *brothel*, the *computer*, *Newton's first law of motion*, and *tactile space*.

However, toward the end of the book there is a small section devoted to "alternate versions," and "chains and clusters" of tetrads for several technologies, including the *airplane,* the *credit card, wine, romanticism, symbolist poetry, visual space, the feminist,* and *rhetoric* (McLuhan and McLuhan, 1988; pp. 195–214). It is in these sections of the book that the real potential of McLuhan's tetrad is hinted at. The essential point is that all technologies always have multiple and sometimes even canceling effects across different arenas or levels of human action and experience. The tetrad also fills out the story of any technological artifact.

When any given device or system comes to market what we typically see and hear about first in the advertising and promotional campaigns are the myriad enhancements it will provide. We also might learn about the impending obsolescence of certain current or previous technologies or practices in the face of the new. We might even hear something about how this newfangled thing brings something valuable back from the past. This nostalgic function may be superficial or significant, but some harkening back to what has come before is common—if only to render an aesthetic/emotional appeal. Unfortunately, we very rarely hear anything about what new sorts of problems arise with the implementation of the innovation. This is the reversal quadrant: what sorts of proclivities or embedded logics of action are bound up in the device, or what might happen if we become too reliant upon the thing, or if it gets overused or abused.

For instance, when portable GPS technology became available we learned how it will enhance our ability to find the way in unknown locales. As can be found in the promotional content for any new technology, we found the immediate benefits being touted, along with information about the doing away of certain bottlenecks or shortcomings associated with some extant technology, artifact or application that had an analogous function. So, in enhancing our ability to, presumably, keep our eyes on the right road, GPS would also render obsolete the traditional fold-out map as well as the practice of having to glance at or thumb through a cumbersome, torn and/or stained, or potentially outdated road atlas. The GPS would also keep us moving. No more pulling over and asking for directions from some unscrupulous (or worse, malicious) person on the side of the road. However the story of obsolescence is almost always woefully incomplete. To be sure, one of the other things we find going the way of the Dodo as GPS technology disseminates is our own, on board, embodied, biological memory. In very much the same way our intuitive ability to recall telephone numbers is inversely proportional to the adoption of mobile phones, my guess is that the driver's intuitive ability to know where he is, or where he is going—and where he's been before—tends to decline with the adoption of GPS. This is not to mention the serendipitous connections made, or special insight gained, during a pull-over with the tip off to that secret swimming hole, flea market or sandwich joint.

And so then to fill out this tentative tetrad, what GPS might retrieve from the past is the immediate feeling of at-homeness, of being native to a place, the sense of knowing one's environs, that small town, country sensibility. GPS allows us, in theory then, to find any place anytime. It simplifies the terrain, subdues the city, thins the forest, and flattens the mountainside. Finally, even though *Garmin, TomTom,* or *Megellan* might not let buyer beware, in addition to a loss of intuitive navigation, the small screens on these devices end up redirecting the driver's gaze like a moth to a flame. In actual use they begin to obsolesce peripheral vision, the tendency to glance around and look at the cityscape or countryside. GPS-induced traffic accidents are becoming more than just a statistical outlier in the NTSB database, and the factory installed systems notwithstanding, these "satnav systems" are also becoming a highly sought-after commodity good for hawk and trade on the black market. But that is just a quick take on what's possible with such an analysis. From these several elements (and a few more to consider) a GPS tetrad drawn up in the standard graphic might look something like the figure on page 34.

What such an analysis suggests is that, instead of seeking clear answers within exclusive boundaries and void of contradiction, it might be better to identify the recurring sets of questions that unify an apparently disparate collection of ideas. However, the tetrad does allow us to garner at least tentative answers to many important questions regarding the implementation of technology. And as I'll illustrate in each chapter, and then again in a nutshell with the corresponding tetrads contained in the appendix of this book, these are always educated guesses about moving targets. After all, as Ruth and Elihu Katz put it in their contribution to a special issue of the *Canadian Journal of Communication* devoted to McLuhan, using his ideas, let alone figuring out what he meant, "is not easy."

[I]t is not easy, first of all because it is difficult to read McLuhan. He refuses to hold still. He takes what Walter Benjamin calls "tiger leaps" inside the world of scholarship. He invents language of his own, well spiced with contemporary jargon. He contradicts himself, often intentionally. Every assertion is probably wrong. And he changes his stance—shifting attention from literature to the content of popular culture, and then to the forms in which the content resides; and shifting from the moral outrage of a critical theorist . . . to the ostensibly value-neutral position of a culture historian (Katz and Katz, 1998; 308).

road hazard

another thing to be stolen

dependence

anxiety (what went wrong?)

loss of control/being controlled

surveillance

movement is data

navigation

safety

control

ego, confidence

sense of freedom

awareness of environs

time/fuel savings

punctuality

map reading

paper maps and atlases

sense of place

happenstance/serendipity/exploration/discovery

roadside interactions/community

intuitive movement/
spatial intelligence

memory

peripheral vision

the vista

GPS technology

ENH | REV
RET | OBS

"knowing" one's way

a sense of direction

straightforward routes

the "greasy spoon"
(goes corporate)

Then again, it's not clear to me that McLuhan ever made any real assertions. Or, if he did, it seems unproductive to engage in an exegesis of his work in search of the true, right, or correct reading of any one in particular. In parts or in whole, little is amenable to formal, logical exegesis. And while this lack of closure seemed not to bother McLuhan, it is undoubtedly one of the reasons he received so many uncharitable reviews over the years. To be sure, the modern, western mind in general, and perhaps the modern American mind in particular, seems acutely intolerant of mystery.

So maybe McLuhan's work is best read as an "open work." Perhaps he is polysemous or multiply meaning*full* in the purest sense of these words. Did McLuhan really know what he was *doin'* as they always wondered aloud on Rowan and Martin's *Laugh-in*? While his earlier work certainly feels more like art than scientific argument, I'll suggest once more that there always seemed to me to be a kind of strategic ambiguity in McLuhan's writing, thinking, and speech. So perhaps in this way he knew full well what he was doing. If more artistic and rhetorical in form, McLuhan's thinking can be the first part of an enthymeme, with his reader answering the aphorisms and filling in the punch lines to his probes in their own way. McLuhan's meaning, in other words, is in his use. Or, rather, the use someone, some close and careful perceiver puts him to. Indeed, it's not *how* one reads McLuhan, it is, above all, *that* one reads him. After all, "[a]s an instrument of action, language cannot serve a representative function. Truth is, in William James's happy phrase, what "is better for us to believe," and "the test of truth of propositions is their adequacy to our purposes" (Richard Rorty in Carey, 1998; 80). Perhaps understanding McLuhan in general, and understanding our Digination in particular is about getting at some of those practical truths. This seems to be what McLuhan was all about.

Chapter 3

Indigenous E-mail

Identity Construction at the Oral/Textual Interface

"People have to change, including us Indians."

—Tom Porter/Sakokweniónkwas

I want to begin this chapter by reminding the reader of a sort of truism that has slowly taken hold in communication studies over the last forty years or so. It's the idea that human symbol use is both socially constructed and socially constitutive. This means that, as social constructions, symbols are conventional and arbitrary, bearing no necessary relation to their meaning or uses. It always catches a few of my students off guard to hear that there is nothing necessary about the relationship between words like *dog, stick*, or *bone* and the things they refer to in the world. Nor is there a natural meaning for the "middle finger" or "okay" signs for that matter. By the same token, a deep sun tan, a "muscle car" or an *iPhone* have no essential significance in and of themselves. The meanings of each of these things are always bound up in conventional understandings that spin out of their specific contexts of use. As socially constitutive however, we know that symbol systems also represent environmental conditions, or worlds which people inhabit and which they accept as real or essential, rather than arbitrary or ultimately even groundless.

A number of theorists have pointed to this phenomenon with regard to media utilization: people who make use of different symbolic technologies not only exhibit distinct ways of experiencing life—they may also possess unique modes of cognition as well. For instance, McLuhan (1964), Eisenstein (1979), Ong (1982), and others have posited that people living before Gutenberg's invention experienced a phenomenology or way of being essentially different from that of a modern westerner. If this is true, it is conceivable that

37

contemporaries with different familiarity and exposure to communication technologies may experience similarly contrasting ways of seeing, thinking, and experiencing. Indeed, we should expect that individuals living in different circumstances than ourselves—be it different landscapes, climates, population densities, or degrees of immersion in modern amenities—may not only develop different means and levels of abstraction in their symbolization, but also experience fundamentally distinct ways of *being* in the world. Positing the existence of such distinct ways of being does not necessitate that we invest in any kind of essentialist doctrine. On the contrary, it asks that we merely recognize the potential any environment has to color our experience, our modes of expression, and even our means of self-conception.

In this chapter I attempt to shed light on some of these interactive dynamics as they occurred between humans and a fairly mundane technological system. I interrogate a Native American group's use of computers in general, and e-mail in particular. The original research was conducted for my doctoral dissertation in the late 1990s, which was based on eighteen months of ethnographic fieldwork. After briefly describing the "media history" of the Iroquois I'll review an abbreviated list of contemporary media research projects most relevant to that history. I'll then present some of the data I collected while working with a small group of Mohawk Indians (one of six tribes that make up the Iroquois Confederacy) as they developed a series of protocols to help guide their interactions with the now ubiquitous digital communication technology that is e-mail.

While the Mohawk have one of the most fascinating (and tragic) technological histories, the primary reason I want to begin the empirical portion of this book with a group of Mohawk using e-mail (or e-mail-using-Mohawk) is due to the fairly rapid changes the community experienced with respect to the introduction and utilization of mediating technologies (including things like mechanized harvesting equipment, VCRs, and a small collection of software applications). I think it is an ideal case, and context, to begin investigating the many issues surrounding human-machine interfaces, and the interface between human culture and media culture in particular. The tension that has long endured between technology-oriented and more humanistic or cultural orientations to technology use can be a useful tension. I think it can help shed significant light upon questions of identity, values, intentionality, technological-determinism, self-determination, as well as the overarching question of autonomy in both humans and machines. It is this tension between two competing theoretical positions, as well as the tension between technology and human *being* that prompted my investigations into the uses of different digital communication technologies by others, elsewhere. This dual tension is part of what inspired the term *Digination*, and is ultimately what motivated me to write this book.

In 1993 a small group of self-described Traditional Mohawk Indians left their reservation on the US-Canadian border with the hope of re-establishing traditional patterns of behavior without fear of persecution. Three years after re-inhabiting the site in the Mohawk River Valley of central New York State (which happens to sit upon the sacred grounds of an original Mohawk settlement), a rudimentary Web page was designed and administered by a non-Mohawk friend of the community at a remote location about 130 miles southeast of the settlement. The new Web address quickly became over-run with e-mail messages from around the world offering volunteer, moral, and monetary support. At that point, the prospect of local e-mail access began to look like a good idea to several community members responsible for funding initiatives. The technology was also being talked about as a good way to maintain correspondence with friends and family. Six months after the remote site went online in November of 1996, several members of the group began using e-mail, a new, and very unfamiliar communication medium— something that was, for them, so clearly *not* them (intellectually, spiritually, culturally, etc.).

The group experienced a series of profound insights and ironies while attempting to reconstruct[1] traditional patterns of life in a rural setting amidst a collection of modern tools, but I'll concentrate on their experiences with e-mail. As biologists are prone to say lately, *the action is at the interface*, and in this case it is the interface between oral and electronic communicative cultures that will be analyzed here.

Without watering down my original analysis beyond recognition, I'll just temper some of the more opaque theoretical language and do my best to preserve the good stuff. That is, after some necessary background and framing, I highlight my research participant's thoughts, observations and experiences with e-mail, something that became one of the most problematic communication technologies encountered by the group.

After a couple months of what really amounted to hard labor, I was initially allowed to take just brief respites from my work in the corn fields, paddock, barn, and tractor garage. Once inside the residence buildings I soon began to hear participants describing a variety of ways in which their thoughts, their perception of self and other, and their typical habits of communication were challenged, distorted, or otherwise perturbed during interactions with the e-mail system (MacDougall, 1999, 2001, 2009).

The initial plan was to study their use of digital communication technology employing as inductive an approach as possible. This included their use of computers, the Internet and e-mail[2], and how they perceived, reacted to and dealt with the various physical and symbolic interfaces between human and machine. As the research progressed I became particularly fascinated with their e-mail activity because of the way they reacted to this application of the computer.

As an inquiry into the relationship between human actors and the particular physical and symbolic environments they encounter, my framing is similar to that taken up by William Cronon in his seminal *Changes in the Land: Indians, Colonists, and the Ecology of New England* (1983). Cronon changed the terms of analysis regarding the relationship between humans and the environments they inhabit from "man *in* nature, or man *against* nature, to man *and* nature." This subtle shift had two important methodological consequences contributing to a much more nuanced understanding of the early contact period. First, Cronon removed Indians from what really is a Euro-centric, historically-imposed role of natural surrogate, to the role of co-participant with Europeans in the complex unfolding ecosystem in which both were involved. This recast Native Americans, not just Europeans, as actors capable of changing the environments and ecosystems with which they mingled. Second, Cronon's work emphasizes the reciprocal, mutually altering dynamic that underlies any *system* (natural or human-made) and, as I interpret him, even suggests an eventual blurring of the distinction between Technology/Civilization and Nature. That is, Cronon describes how not only the land, but the Indian too, was significantly altered by wider changes in the surrounding, total environment.

Likewise, as things progressed at the Mohawk settlement, the reciprocal relationship between my research participants and the new technological environment they were exploring became impossible to ignore. Mohawk cultural history also happens to be a very rich and complicated media history. So before we get to the details of their interactions with e-mail let's consider some of the history that led up to these encounters.

A (BRIEF) MEDIA HISTORY OF THE MOHAWK: COLONIZATION, "CULTURAL SCHOOLS," AND NEW TECHNOLOGY

When it became clear that I would be granted access to the back offices and residential spaces where computers were installed at the community, I began some fairly in-depth historical research on the Iroquois tribes that continue to live in and around what is now New York State. These investigations of Mohawk and other Iroquois' experiences around the time of sustained European contact in the middle of the seventeenth century suggested that there might be an element of *media determinism* at work throughout the population's history (cf., Cook, 1989; Snow, 1994; Snow, Gehring and Starna, 1996; Boldt, 1993; Mander, 1991, Porter, 1997; Vachon, 1991).

This academic research gave me a solid foundation, of course I learned so much more from Tom, the leader of the community. The Mohawk's oral tradition is replete with stories describing the introduction of foreign "media" (e.g., languages, alphabets, printed texts, music, alcohol, etc.). Indigenous people in the United States have been talking for a long time about the various ways they have been impinged upon by what is variously described as European, Western, or "White" technology and culture. They have been talking to each other about such things for centuries now.

There is a clear but subtle theme woven into the oral and written record of the Mohawk. It is a theme of resistant yet progressive compliance to a variety of non-Native modes of expression. As the European colonial influx continued throughout the eighteenth and nineteenth centuries, many Native groups found themselves confronted with basic questions of survival. Most of the stories I heard from Tom and other elders and community members suggest that the Iroquois leaders were always aware, since the outsiders were here to stay, that several elements of the European world would have to be integrated into their own. Without certain things, whether it was the more effective firearm of the fur trader, the antibody for smallpox that most Europeans carried, or the languages which enabled the dissemination of diplomacy and commerce, indigenous people would need to incorporate these elements to varying degrees, or risk almost certain extinction.

One particularly poignant bit of folklore tells of advice from an Iroquois elder to his clan who finally admitted, upon seeing how smallpox was decimating his people, that the time for running from the white man was finished, and that the time had now come for some of them to lay with the intruder instead.

Tom extended this theme of compliance—of "adaptation without assimilation" as he put it—during a number of conversations with me as it related directly to language and technology. Tom spoke about these issues in connection with a central aim of the community, the creation of a Mohawk language and cultural school. The plan extended the hopes of the group, including the return to traditional patterns of thought and behavior. These patterns, these ways of thinking, seeing, and listening continued to be invoked within the context of a wide range of telecommunication technologies, including telephones and fax machines, radio, satellite television, audio and video recordings, and computers (via e-mail, some record-keeping programs, and sporadic Internet searches). Despite these ironies, however, the plan was clear: under Tom's careful guidance the group would return to service as many of the cultural practices and values as possible that had been lost in and through the Carlisle Indian Boarding School system. In essence, then, spoken language was seen as the defining, indeed the *self*-defining, constitutive medium through which all else was filtered. Consider the following excerpt authored by Tom that was featured on the community's original remote Web page:

> The Iroquois nations are at the edge of a cliff in terms of becoming extinct— linguistically, culturally, spiritually, morally and politically. The only thing that will remain are biological Iroquois, but nothing else.
>
> The Carlisle School[3] and other schools like it were the main culprits in what could be the potential demise of our people. They took away our language, our family structure and connections so that those affected don't know how to be parents and grandparents to our children. They took away our religious and spiritual beliefs and, in so doing, took away our culture.
>
> All these elements combined caused us to become a dysfunctional people right down the line so that those of us here today are dysfunctional as well. As a matter of fact, I really don't know of a single healthy Indian family in the entire country of America or Canada.
>
> Now as a possible future project, the dream, if possible, is to make our very own Carlisle Indian Boarding School except everything will be in reverse. All the things that Carlisle took away, we will attempt to return to our people.

It was Tom's contention that if something was not done quickly there would be only "empty Mohawk bodies left walking through the world," with "dead eyes" and "cold spirits." Tom described how Iroquois' modes of perception and expression were almost entirely wiped out with the introduction of the Dutch, French, and then English in the mid-seventeenth century, the incorporation of a *lingua franca* consisting of these three, and the eventual adoption of the continental dialects in the region. In Tom's opinion, most of what is essential and pure in the Mohawk people resides in their language, which, in turn, houses the very basis, the "heart and soul" of their culture.

A similar account of this past was relayed to me by another informant who visited the community during festivals. He spoke of a number of artifacts and practices (dubbed "mind changers") introduced by European explorers and settlers that wrought havoc on all Native groups encountered. These included alcohol, the Bible, playing cards, and the violin. Concerning the first three, Tom suggests that one merely turn their eyes toward any reservation to see the horrific consequences of cultural assimilation. As for the violin, this instrument came to symbolize the decadence and superfluous nature of European society. Thought to deprive the Indian of his ancient rhythms, it emphasized the flighty melodies and contrived world of the white man.

It has also been suggested by numerous researchers that the incremental loss of language was a primary factor in many Native persons' assimilation into the media-rich aspects of mainstream American life (Gehring and Starna, 1996; Boldt, 1993; Mander, 1991, etc.), though most did so without any of the socioeconomic benefits often associated with that mode of life. My research into the history of

Iroquois language and culture surrounding the lengthy contact period spanning the seventeenth, eighteenth, and nineteenth centuries suggests the existence of a mode of experience that was fundamentally different from that fostered by the European languages and culture that has come down to us since that time.

In our personal conversations, and during several public talks at public festivals, Tom lamented the toll exacted upon his own upbringing by the English language. For example, he attributed his parents' difficulty in displaying affection to him as a child to the very forms and patterns of speech and social interaction enforced by these English-only cultural de-programming/re-programming schools. "Carlisle Schools," is the generic name for a legion of similar facilities built throughout the nineteenth and early twentieth centuries advocating systematic, institutionalized programs of "humanitarian aid" for American Indians (Pratt, 1908). To be sure, many Mohawk I encountered during my research believed that such schools were being established well into the twentieth century, and that many still exist even today under different names and guises. According to Brigadier General Richard Pratt, the founder of the original Carlisle School, the task was to salvage what humanity remained in the savage. The unofficial motto of that first institution in Pennsylvania was to *"Kill the Indian and Save the Man."* The biological metaphor is easy enough to detect here, as is the impression that a tenacious pathogen is the culprit. That pathogen, of course, would be the language-qua-culture of the Indian.

Pratt concluded that civilization was the answer to the Indians' problems. The idea was to "[c]onvert him (the Indian) in all ways but color into a white man and in fact the Indian would be exterminated, but humanely, and as beneficiary of the greatest gift at the command of the white man: his own civilization" (Utley, in Pratt, 1908, p. 5, emphases in original). The idea was that the patterns and means of expression, ways of speaking, seeing, hearing, and thinking, cultivated by the English language would replace the Iroquois' natural, less refined, more savage means. The plan was to rid the "student" of all essential Indian properties and characteristics, short of skin tone and other phenotypical aspects common to indigenous groups in the region. Pratt told a Baptist convention in 1883, "I believe in immersing the Indians in our civilization and when we get them under, holding them there until they are thoroughly soaked" (Pratt, 1908; p. 5). Interestingly enough, the widespread dissolution (and forced replacement) of their language is usually the first problem or "sickness" cited by community members, and their commentary forms a collective vision describing nothing less than an all-out war of symbolization and representation that has ensued for centuries.

The second, though somewhat simultaneously experienced medium, was actually a set of intrusions into the Iroquois' world, namely: literacy, writing, and all the habits of mind associated with them. Early attempts to assimilate aboriginal people in at least the northeastern regions of what is now the United States included the invention and imposition of an abbreviated alphabet by Jesuit missionaries to adapt a number of Iroquoian languages. "The Jesuits gave them the Alphabet to help them parse out their thought. The French and British gave them the notion of property to help them parse out their land" (Vachon, 1991; p.16). Of course, these impositions were exactly that: forced adoptions of alien communication systems. These new symbolic phenomena and forms positioned the Natives not only in relation to dominating groups and their different modes of expression, but also in relation to themselves. Both the historical literature and my early discussions with informants at the community suggest Iroquois groups involved in first contact had a general awareness of the powerful role played by certain media in this process.

Several of the Mohawk I talked with about this mentioned a twelve-character alphabet that was imposed by the Jesuits nearly 350 years ago. Other linguists and historians (some Mohawk, some not) vary this number slightly. What seems clear is that the Mohawk and other Iroquois groups were involved in a cognitive/ phenomenological shift that was prompted in part by a process of *noun-intrusion*, but perhaps most directly by two even more systemic forms of cognitive and perceptual change: the increased abstraction and decontextualization that comes along with the conversion, alteration and re-representation of their language through chirographic and typographic technologies.

While numerous proverbs[4] dating back centuries hint at the basic idea, the first formal treatments suggesting that different languages tend to foster different kinds of subjective and collective experience (which, in turn, lead potentially to distinct life-worlds, or ways of seeing and being in the world) can be found in German philosophical thought of the late eighteenth and early nineteenth centuries. The intellectual lineage certainly includes the perceptual categories of Immanuel Kant, and the anthropological work of Franz Boas. In the twentieth century, a much more formal statement regarding a kind of linguistic determinism can be found in the work of Edward Sapir and Benjamin Whorf. Consider the way Whorf describes aspects of the *Sapir-Whorf hypothesis*:

> We dissect nature along lines laid down by our native language. The categories and types that we isolate from the world of phenomena we do not find there because they stare every observer in the face; on the contrary, the world is pre- sented in a kaleidoscope flux of impressions which has to be organized by our minds—and this means largely by the linguistic systems of our minds. We cut nature up, organize it into concepts, and ascribe significances as we do, largely because we are parties to an agreement to organize it in this way—an agreement that holds throughout our speech community and is codified in the patterns of our language (Whorf, 1956).

The notion that writing and reading might additionally alter both subjective and collective experience gained acceptance in academic circles during the second half of the twentieth century. This more causal or "formative" view of communication media amassed considerable support in the discipline as it relates to the social effects and functions of various modern technologies including mass-produced print, radio, television, computers, and the Internet (Eisenstein, 1979; Goody, 1959, 1969, 1977; Gumpert and Cathcart, 1982, 1985; Innis, 1950, 1951; McLuhan, 1962, 1964; Meyrowitz, 1985; Ong, 1982; Postman, 1985, 1992; Strate, Jacobson and Gibson, 2003).

Tom guessed that "something like two-thirds of a people's culture is bound up in their language." Clearly for him, the language contains a great deal of formative power and energy. "The way we talk," he continued, "paints a picture of the world that is very ancient, very sacred, and very special. For some Native groups there are only one or two dozen native speakers remaining. When they die, that way of seeing the world will die."

In the months following the installation of the e-mail system several members of the community began using the technology to communicate with friends in neighboring communities, as well as with strangers around the globe. During the course of my fieldwork, Tom and others complained about certain aspects of their e-mail system. Most of these folks appeared to be having trouble with specific features of the interface—like the subject heading and user prompts. There seemed to be a disconnect between oral language, far and away their primary mode of communication, and the demands placed on them by the machine. I became very interested with the idea of illuminating some of the ways the medium potentially enhances or inhibits the transmission and reception of symbolic information (of communication, culture, and subjective experience itself) from the standpoint of these new technology users.

Before we consider some actual e-mail messages produced by these users, I'll need the reader to be patient for a short while longer. Users accounts of their interaction with the technology are interesting in and of themselves, but it will very useful to review some of the theory that informed the larger project to show how these users seemed to have intuitive knowledge that sometimes corresponded quite nicely with the seminal theories.

ON PRIMARY AND SECONDARY MEDIA

There are several key distinctions made by proponents of media ecology that will be useful to us here. They are distinctions finally gaining legitimacy in other areas of communication and technology studies as well. A central idea is that some communication media are primary, whereas others are secondary. Media are primary in the sense that their grammars or rules of use are learned early in life, through tacit socialization. Certainly spoken language qualifies here. Whether or not one subscribes to a Chomskian theory of innate grammar, we seem to acquire language in a sort of perfunctory fashion—that is, in a sort of mindless, or pre-conscious way. We learn to speak without even trying. This is the nature of a primary medium. We tend not to see it, or think about it, in any conscious way. If some media are primary it follows that other media are secondary. While the time-lags vary, secondary media are often learned later in life through more explicit instruction and with some dependency on, or filtering through, the primary medium (Whorf, 1956; Birdwhistell, 1970; Sapir, 1983; Gumpert and Cathcart, 1985).

Now, it is also important to point out that these two terms—primary and secondary media—are not absolute. A technology such as the graphical computer interface, can be a primary medium for one person (like my 14 year old nephew), and a secondary medium for another (usually, but not necessarily older) person, like his forty-year-old father.

Due to early and sustained interaction with extended family members, Tom and several of the older Mohawk who participated in this study consider the Native dialect to be their primary medium. However, all of the participants experienced the world through a primary medium in which they were enmeshed and which theoretically structures and directs their cognitive processes in certain ways. That medium is their preferred and predominant mode of interaction—namely, face-to-face talk, no matter the language or dialect. One reason we might detect problems or incongruences between their talk and their e-mail use is that the technology is functioning as a secondary medium for these people in a way that is distinct from its use for most of us in the mainstream. E-mail is interfacing with the primary medium (face-to-face, or F2F talk) and creating interpretive problems for these users. I thought it was conceivable, in other words, that due to their particularly deep level of enmeshment in oral communicative patterns and processes that this new interface was triggering symbolic perturbations within and between the rule systems or underlying grammatical structures of the respective media. F2F talk and e-mail are certainly not the same kinds of things.

This distinction between primary and secondary media suggests that the experience of any medium is not derived from the medium itself but, rather, the relative sequence in which one gains fluency in a medium (a unique symbol system and signaling system) that more reliably accounts for the experiences one has with that particular medium (c.f. especially Gumpert and Cathcart, 1985). This suggests, in turn, that it is a relative, not absolute, distinction. Thus, for many children today, television and even the Internet are primary media. Counterintuitively perhaps, and if the proper conditions allow, many people can experience and begin to learn the structure of those media (visual/iconic literacy), before they learn the structure of the oral/linguistic concomitant. In other words, for the first time in human history we are seeing large numbers of people initially orienting themselves to the world in a way that is very different from the way the human race evolved.

While this is not the primary reason the American Pediatric Association made a recommendation at the end of the millennium advising parents to keep tabs on their children's use of certain electronic media (like prohibiting television viewing before a child has reached two years of age), part of the concern was based upon research showing that language acquisition can be stymied by over-exposure to visual media (Howard, 2004; pp.362–63)

At the same time, however, we can begin to understand how, for someone like my mother who was born in 1930 and reared on print and radio, television can be a distinctly secondary medium. Indeed, for many people alive today, shows like *Survivor, Lost,* and *NCSI* are often very confusing presentations of sound and action issuing from the television screen. A similar problem parsing information even shows up when some Europeans first experience American television in its native form. Here in the States, the distinction between the program and the advertisement is often very subtle and can easily be missed. In France, by contrast, when a program has finished and a series of advertisements begins, the transition is marked clearly with the notice *Publicite'* appearing in the lower right hand corner of the screen. Or take Britain, where, through the 1990s there were still no ads on the four public channels. In 2004, however, Britain became the first nation in the world to set up the first channel dedicated entirely to advertisements. In these senses, then, there seems to be a functional difference, and some recognition of a *difference in kind* manifesting between media forms and formats that might often be misconstrued as the same medium.

However, it is not necessarily by virtue of my research participants being Mohawk that e-mail is a secondary medium for them. Rather, it is the fact of their being Mohawk firmly rooted in a predominantly oral, traditionalist, rural enclave that adds credence to this distinction. Their own experience with various media is what most readily determines their relationship to this particular technology. The perspective offered by Bavelas et al. (1997) is worth considering in this regard. Using face-to-face talk as their baseline, these theorists suggest how other communication media might be impoverished in various ways. For example, they suggest that: "[t]he natural origins of communication are in face-to-face dialogue. Dialogue is not only our first form of communication, it remains our most common in everyday life . . . face-to-face dialogue has features that are often missing or highly limited in other forms of communication—hence our suggestion that dialogue might serve as an appropriate standard for other communication systems" (p. 23). If this assessment regarding the prevalence of F2F dialogue applies less well to many people's lived experience today, the evolutionary significance attributed to that mode of communication still holds.

Thus, given the emphasis so many people place on the efficiency and pragmatics of communication in this fast-paced world of electronic interchange, there may be some problems with such an assessment. But even if it is questionable whether face-to-face talk should still be considered an appropriate yardstick for assessing all other communication channels, my fieldwork experiences with the Mohawk led me to believe, at least for my research participants, that such a comparison was intuitively taking place. The problems my Mohawk friends were having with e-mail seemed to derive in large measure from their cultural assumptions about how communication operates, which in turn derived from the importance of the face-to-face context using oral language.

So beyond (or in addition to) Bavelas and associates' concern for an ideal communicative type, we need to be looking at the interfaces between different forms of communication (i.e., symbolic forms) and the way symbolic exchange at the interfaces between these forms alter the representational performance of various media. In other words, certain communicative tools might be better suited to certain communicative tasks.

There has been a lot of speculation in this chapter so far, and media researchers can certainly be prone to speculation. This is why I think these kinds of questions about communication technology are best answered from the vantage point of the individuals actually doing the communicating. I was interested in whatever incongruences arise when these technology users confront a new symbolic medium, as well as any ways the medium enables and/or constrains their usual interpretive activities. By "incongruences" I point to any perceived problems with the technology, something which prompts a user to pause and maybe even question what is going on. The problem, in turn, often leads to a felt need to alter the "conventional" use (and, subsequently, the very structure and meaning) of the medium.

To sum up the present discussion, human communicators encode content (symbols) in unique ways by selecting the various media at their disposal to hopefully "get the message across." However all media encode content in unique ways.

EARLIER RESEARCH IN CMC IN AND OUT OF ORGANIZATIONS

Some earlier computer mediated communication (CMC) research in organizational settings recognized the importance of communicators' practical goals and so looked to the user/communicator as the appropriate unit of analysis. A collection of these scholars working from a user-orientation in media studies acknowledged the *plasticity* of individuals with respect to their ability to make up for the narrow symbolic capacity attributed to various media. For instance, even if Avery and McCain (1982) contend that "media-personal" encounters are inherently different than face-to-face or interpersonal encounters, they do not feel as though these differences are as deterministic or as prohibitive as one might suppose. While the authors did not address CMC specifically, their insights are notable inasmuch as they acknowledge the basic differences between F2F and mediated encounters to be a difference in sensory potential (with the key word being *potential*). Avery and McCain rest their case on differences found in sensory potential (number of senses and other perceptual modalities being activated), control over the exchange, and knowledge of the source.

On this view, and quite contrary to Bavelas at al., face-to-face communication becomes little more than an arbitrary baseline that does not and has never necessarily implied full sensory integration. Rather, such integration, and the experience of "real communication" is less a matter of the specific features of various media than personal interest and participation. This set of assumptions, along with much more serious attention toward promoting bias-free investigative contexts, has allowed for some very interesting, if at times, counterintuitive, findings of user-based studies of computer-mediated communication (Bench-Capon and McEnery, 1989; Boczkowski 1999; Walther, 1994, 1996).

For instance, Walther (1996) poses a very good question: "How can the same group of technologies be described as a limitation and a hindrance?" (p. 3). Walther's question might be restated: If media are stable in terms of their ability or potential to relay and convey content, then why do specific media appear to perform differently in different situations? Part of the answer, as Avery and McCain (1982) intimated early on in their study of interpersonal telecommunication rests with the individual(s) involved. One reason why media are perceived to be differentially enabling is because the intentions, needs, desires, and relative interest and attention paid by media users differ across studies. This seems to be why Kerr and Hiltz (1982) also reported that individuals using early text-based conference and messaging systems adapted to the medium rather quickly, so that degrees of "impersonalness" eventually reduced as well.

In similar fashion, Myers (1987) suggests that, through the in situ development of symbolic meaning, users of electronic bulletin board systems (BBS) were able to maintain their identities as leaders. Myers posits that they were able to do this by actively manipulating the computer-mediated context. Similarly, Feenberg (1993) provided an empirical account of a kind of reinvention process originating in user's often idiosyncratic (re)interpretation of a technology subsequent to its introduction into a social milieu. These studies predicted some of the phenomena I observed at the Mohawk community. However, there are several additional ways to frame the Mohawk/e-mail situation.

One way to think of a group of technology users is as an "interpretive community." This term, coined by Stanley Fish (1980), originally referred to the sub-groupings that develop around different readings of the same written text. A similar articulation of this idea can be found in Radway (1996). Both of these works, however, deal exclusively with the multivalance of a written record. Lum (1996) introduced a new dimension which widened the occurrence of interpretive communities, perhaps even making them as ubiquitous as the perceptions of the communication modes themselves. In that study, Lum suggested that the practice of Karaoke (and all of the audiovisual technology supporting that practice) resulted in an array of interpretive groups. Likewise, Umble (1996) conducted a research project around the use of the telephone in traditional Mennonite and Amish settlements. My thinking was that there may be similar interpretive processes occurring at the Mohawk settlement.

Additional research along these lines includes the work of Carbaugh (1988, 1990) wherein ethnographic approaches reveal some of the many ways Americans position themselves culturally through the kind of talk they engage in with others. Weider and Pratt (1990) hone this approach and focus upon methods and means of identity construction among American Indians. In that study, representatives from the Osage tribe (Pratt's own) were shown to invoke more general features of "Indianness." The authors discuss a kind of *cultural repertoire* that has been established over time to which Indians tacitly refer in order to be Indian at appropriate socio-temporal junctures. The difference with the present inquiry is that these Mohawk seemed to have no such repertoire to borrow from during interactions with (and through) e-mail. Whatever new grammars and logics came along with the medium, the technology was also an, as yet, undefined or uncoded artifact for them. They were still working out the meaning of the medium and seemed to be attempting to do so from a cultural vantage. With this cultural positioning notion in mind, let's now consider some of my Mohawk friends' uses and interpretations of e-mail.

One of the first observations made by my research participants concerned the problem of categorization within a subject heading. That initial "demand feature" of e-mail was perceived as constraining to users in a number of ways. The most common complaint dealt with the struggle to contain meanings, or "gists," in a single term or phrase. This seemed plausible enough. The subject field is a formal feature of the e-mail application that calls for the categorization of messages within a header. It was a demand treated with substantial trepidation during the first several months of e-mail activity at the settlement. I wondered why this reported difficulty caused these users so much trouble. One idea was that in keeping with users' verbal utterances, their outward behavior seemed not to reflect underlying categorical processes (i.e., sequential/categorical logic, pre-formed decision trees or other associated modes of cognition considered "hierarchical" in nature).

SOME REMARKS ON THEORY AND METHOD

If one looks closely, they'll find that the majority of investigations into modern media use accomplished over the last several decades (since such formal study essentially began) fall into two categories. That is, they tend to proceed both theoretically and methodologically along one of two ways—the *medium* or *cultural/interactionist* approach. The medium theoretic path, which often makes use of artificial experimental contexts designed to detect the manner in which users "give in" to the various stated *demands*, *logics*, or *representational shortcomings* of the medium, tends to ignore many of the novel and creative ways in which users employ the medium, as well as any of the associated practical implications such employment holds for the users themselves (see in particular, Culnan and Markus, 1987; DeSanctis and Gallupe, 1987; Kiesler, Siegel, and McGuire, 1984; Sproull and Kiesler, 1986).

The cultural/interactionist approach can also be limiting in and of itself, as that method often leads the researcher to look for any group-based interpretations and meanings associated with the medium that result from new uses and functions associated with it. The danger here is in neglecting or devaluing behaviors that might be influenced more directly by some formal, structural, even universal feature of the medium itself (see in particular, Bench-Capon and McEnery, 1989; Lea and Spears, 1995; Rice and Love, 1987).

But while many investigations into computer use have proceeded in such discrete fashions, there are ways of integrating the medium and cultural/interactionist approaches so as to produce more comprehensive analyses. Indeed, hybrid methodologies may outperform more "pure" methods in a variety of ways. A number of precedents exist for such an integrated approach. In these cases the researcher remains open to the possibility of multiple factors coming into play simultaneously (e.g., the medium "acting on" the user and the user "acting on" the medium). These are occasions when subtle interactions between a medium's features and culturally specific uses are identifiable (see especially Boczkowski, 1999; Lum, 1996; Umble, 1996; Walther, 1994).[5]

A wider interpretive frame is one significant virtue of studies that work to integrate both perspectives. For instance, this is one of the enduring contributions Lum (1996) makes generally to the field of media studies. Lum's investigation of karaoke technology by a group of Chinese Americans living in the New York metropolitan area suggests that medium-based demands can and often do exist in conjunction with culturally specific uses of the medium. Lum reveals how people construct, maintain, and/or transform their social realities and their attendant conception of self vis-à-vis any number of extant artifacts. He suggests that the way individuals' engage with various technologies "can at the same time redefine the nature of their social experience" (p. 106). Ultimately, however, Lum recognizes more of a bidirectional interplay between medium and user—or technology and culture—than many investigations have been able to do (due mainly to limitations in their methodological approach).

The present investigation of e-mail use (like most of the chapters in this book) tends to follow Lum's lead by employing both medium and user-based interpretive frames. To reiterate, while medium theory and cultural/interactionist theory each explain discrete pieces of the data at hand, I'll show, like Lum, that they can also exist in a complementary relationship. For example, we'll see how the structural features of the Netscape (v.3) e-mail interface altered, and was altered by, habits established by individuals using the system. I think this dual approach permits an enhanced picture of events surrounding e-mail use at the Mohawk community. Such a dual approach can help us get our collective head around what is going on with a lot of other technologies we find ourselves immersed in too. Finally, such an approach also helps clarify what I mean when I say that Digination is a reciprocal process, and even a sort of biological phenomenon.

Given the emergent nature of my investigations with the Mohawk, this chapter ends up having less to do with the technical features of computers and e-mail use per se, and more to do with the various ways in which this small community of users could be observed justifying their employment of a technology—to each other, to a university researcher, and to the mainstream world. That is to say, while observations filtered through the medium-theoretic and cultural lenses can tell us something about the interface of electronic and oral communicative logics (i.e., grammars, biases, etc.), and their relative resistance to change, there is also substantial light shed upon the way participants invoke, create, and sustain community-based meaning, or otherwise position themselves culturally through their interaction with a new technological artifact.

On occasion I observed what appeared to be wholly unique uses, interpretations and explanations of e-mail (most notably, those interactions surrounding the subject field, that one-line dialogue box at the top of the screen when a message is opened in any e-mail system). For instance, conversations with participants regarding their tendency to balk at the subject field suggests there was a felt pressure to make some kind of succinct explanation, or offer an explicit, clear description of each message at the outset before anything had been typed. Typing in a subject as a topical identifier was generally thought of in the following ways: "That's not right. . . we don't do things this way," "You don't do this first. . . well, I wouldn't do this," and "This isn't normal. What I mean mostly is that it's not natural."

There seemed to be an *a priori*[6] aspect to my research participants' interaction with the medium. There seemed to be something rooted in them which prompted a kind of reflexive response when the medium "demands" things. For example, I noticed them expecting certain kinds of information in the subject field, content that reliably explained the information that was to follow. In other cases participants offered detailed explanations concerning what it means to interact with the medium of e-mail as a Mohawk, even a Traditional Mohawk user. But while I do think something legitimate was going on there, I also figured, given the way Native Americans have been hailed in the media and in pop-cultural texts through time, that they should also be granted some latitude in their self-attributions. In other words, it was likely that an emergent "natural" inclination might start manifesting itself among these Mohawk users that would attribute certain features of their use of e-mail to their own special "otherness." Was their experience of the technology rooted in their *"Mohawkness"?* If so, what might this really mean? Did it imply some essential way of being encoded in their DNA? Or, was their "otherness" more a side effect of their general naïveté about the technology? In an effort to keep a handle on these kinds of questions I established an ongoing correspondence with a small collection of e-mail users off-site (in "the main" so to speak) to compare with the Mohawk data. However, while some naïve mainstream users did also balk at the subject field, their subsequent explanations did not include "cultural" explanations for having done so.

There was a curiously ironic quality to the cultural explanations offered by my Mohawk informants. Most went to considerable lengths making sure that I understood how e-mail, including all of the behaviors that surround e-mail use, is not, and has never been part of the established cultural repertoire of Traditional Mohawk people. Nonetheless, they also seemed willing to present their behavior as *evidence* for a "Mohawk way" of using e-mail. Yet, aside from one research participant—twenty-two-year-old Aha, who had used e-mail several times prior to taking up residence at the settlement—no one had any experience with the medium whatsoever. In other words, my informants responded culturally to the medium. They oriented to it as if influenced by their culture even though they appeared to have no *a priori* cultural repertoire (i.e., set of rules or knowledge-base) that would permit them to define and/or orient themselves in such a manner. Consider the following message text, and a portion of interview transcript that directly references the same message:

>HI TOM IT IS BOBBY WE MADE IT BACK HOME SAFELY WE GOT 8 DEER >I WISH WE COULD OF BROUGHT YOU SOME BILLY CUTS MEAT GOOD >HE HAS A SHOP. DUSTIN GOT HIS FIRST DEER HE HAD TO BITE THE HEART. >THANKS FOR LETTING US STAY WITH YOU. EVERYONE HAD A GOOD TIME. ITS >TO SHORT. THE WEATHER IS WARM HERE IT IS HARD TO BELIEVE IT IS SO COLD >THERE AND SO NICE. WOW YOU SHOULD SEE THESE POINTS TOMMY. YOU >KNOW THE BUCK WE GOT LAST WINTER. I HAVE HIS BROTHERS RACK IN MY >LAP RIGHT NOW.

>HERE JUDY WANTS TO SAY HI

>

>Hey thanks for taking care of my husband on Thanksgiving. Making sure
>he got enough turkey. They all had such a great time. We missed him
>but was glad he is getting to know your family better. Well I best
>close. Congratulations on your new grandbabies. They just sat on the cat.
>Love Bobby and Judy and Kids.

Now, consider Judy's explanation during an interview when she and I reread the message together several weeks after she authored her section:

Judy: We always share the talk. Well, on the phone we always do and even though we aren't reeeeally talkin' there [on e-mail] we can make it seem like the same thing . . .

Interviewer: Who do you mean by "we"?

J: Well, me and Bobby for sure, and Tom too and my cousins and—well a bunch of folks.

I: You have messages where Tom was sharing the writing?

J: Oh yeah, he had Tahawitha [his nine-year-old daughter] send a few lines over a few times, and Phil too.

I: Hmmm? Oh, yeah I saw Phil doing that, but I never saw Tahawitha doing that. But all those people you mentioned—are they *all* Mohawk?

J: Yup -No! Gail is Ojibwe, but she married a Mohawk (laughs).

I: The phone seems to make this sort of thing easy. Why do you suppose you do that same kind of thing on e-mail? I mean I don't think it's that common and I keep an eye on a lot of people's e-mail these days.

J: Well, I think that's just our way of keeping the family and all tight. Keep 'em together. That's part of being Iroquois in general too—it's not just us. And we've got a lot of Bear clan, and I believe the Bear clan is known for that—for family. Well, let's see now, Tom's Bear and Bobby's Bear of course. But a few are Wolf, but those Wolves are real close too. Actually they're close and private. So, yeah, on the whole the community really counts for something with us.

I: So do you think I'd find more of this kind of thing with others at the reservation?

J: You know I can't speak for all that. Cause, see, there's a lot of hurt up there. There's a lot of broken families up there. Yes. Many of the men are out working the steel and they come home every two or three weeks or so. That's all. Sometimes not even that. Maybe they *should* try it [e-mail]. At least they'd be tryin'. But I can't talk for them. But we do it—we keep the family tight. And, well yes, I can say that this is what a healthy Mohawk should do. We try and share communication you might say.

Despite participants' self-reports, as the fieldwork progressed I had to acknowledge that some of the behaviors these Mohawk users were engaging in (like the all-capital construction, run-on sentences, the dual-authorship, the real-time referencing, etc.), that they claimed to be cultural in origin, were not so rare, let alone culturally unique. This problematized their explanations for me. Although the community members involved in this project proffered cultural explanations for the ways they used their e-mail system, I was able to observe very similar uses and behaviors elsewhere, among non-Mohawk. What is more, I did not observe the latter, non-Mohawk users explaining their computer-related behavior in cultural terms.

The thirty-two individuals I employed as a kind of mainstream control group used e-mail at two large corporations, at the local State University, and in various domestic/private settings. Again, these users responded to the same early Netscape 3.0 environment in ways very similar to my Mohawk research participants, yet did so with a verbalized perspective that lacked the cultural content. For instance, where a thirty-something *Sony* executive working in New Jersey talked about making use of Netscape's small composition window as a matter of convenience, two of the younger Mohawk men (Will and Aha) both commented upon that feature of the medium as allowing them to abide by their cultural/ethnic predispositions—to be more "true" or "real."

So there are really two conclusions one can draw from these observations, as well as the use of the medium and cultural/interactionist theories, that need further elaboration: (1) the Mohawk proffered cultural explanations for a technology and use of that technology for which they had no prior cultural patterning and (2) they proffered cultural explanations for accommodations to, and uses of, e-mail that are similar to those found among non-Mohawk e-mail users. Two questions then arise: (Q1) what is the status of a cultural explanation/claim absent some basis for demonstrating a cultural routinization associated with that explanation and/or behavior; and (Q2) what is the status of a cultural explanation in the presence of evidence that others engage in behavior without those explanations?

A NEW AND INCONGRUENT EXPERIENCE

Indigenous cultural groups like these Mohawk seeking out traditional modes of existence may represent test-beds for gleaning the role communication technology has in the formation and maintenance of individual, group, and even cultural identity. American society, even with its wide-ranging social strata, is so thoroughly saturated by most modern telecommunication technologies that there is virtually no way of separating out the signal from the noise, or the *figure* from the *ground*, as it were. This should be no revelation. How *do* we define American culture these days without making reference to communication technology of all kinds? Communication media are part and parcel of contemporary mainstream American life. Put another way, for the most part here in the United States at least, *media are culture.*

In fact, given that media and culture are so intimately related in the main and in the margins, an important rationale for the present investigation was the conviction that these media have yet to sufficiently characterize traditional Native culture (Boldt, 1993; Lafitau, 1974; Landsman, 1988; MacNeish, 1952; Mander, 1991; Snow, 1994). My initial working assumption, therefore, was that such cultures are still distinguishable from their technology-based counterparts, and that the particular influence of computer technology on "Traditional Native" cultures could be detected with careful observation and analysis.

It seemed that these Mohawk's communicative practices, more than anyone's I had come across, were firmly rooted in the oral tradition. Given this, e-mail might provide these users with a way to think about their thinking (and themselves) in a way never before possible. Sure enough, after the interview process got underway, our conversations were replete with personal reflections centering upon issues of identity, community, and wider questions of consciousness. All of these emerged during interviews surrounding their technology use. Now, these were not "virgin" technology users. While practicing what they describe to be a traditional Mohawk lifestyle as *People of the Longhouse,* several members of the community were quite literate when my fieldwork began. These folks had even been using typewriters and/ or word processing technology for the majority of their adult lives. Tom, the leader of the group recounted using a DOS-based word processor "in fits and starts" as early as 1984. I located the dusty 8086 IBM PC machine on the premises and had to retrofit my own desktop to accommodate the 5¼ floppy drive so I could inspect some of those early word processor files Tom had produced. Another member of the community, a man in his mid-twenties, had been using word processing programs, spreadsheets, and a graphics workstation including color scanner and fax for nearly five years. But, while significant, the particulars of residents' familiarity with a host of computer technologies are of secondary importance. What should be stressed is that some came to these technologies relatively late in life and utilized them for a variety of distinctly practical ends; ranging from letter writing to keep in touch with family and friends, business transactions, bulk (snail) mailings, as well as the creation of visual art and poetry. The critical point is when the group began using writing in conjunction with near instantaneous electronic communication technology (i.e., fax and e-mail, though e-mail alone is analyzed here). And we will see shortly why e-mail is considered by users to be so new, strange, and different.

Initial observations and interviews revealed that users took notice of what I eventually described as the "linear and sequential" nature of e-mail. They made explicit reference to the way the words follow in succession from left to right and down the screen. For example, despite his experience with first-generation word processors Tom was immediately fascinated by the conjunction of this scrolling feature and the novelty of a fresh message. The ability to scroll up and down a lengthy message just received was perceived to be also different somehow from the *re-readability* of a paper letter. My research participants' descriptions seemed to fix upon the active nature of the visual presentation—how the lines of text themselves appear to move about the surface of the screen.

This, along with the subject heading and its automatic prompts, created two immediate impressions for these users. Upon experiencing the prompts to enter a subject, several described a kind of liveliness or sentience emanating from the machine. This perception, in turn, often led to a straightforward set of responses. Users tended to "satisfy" the machine, as it were. Whether they actually inputted commands based upon a kind of concern over doing what seems desired, almost all observed entered a subject of some kind rather than trying to send off what was described as anything ranging from a "rude," "nameless," potentially "confusing," or even "unworthy" (i.e., illegitimate) message.

The ability to scroll around within the text of a contiguous message also contributed to this perception of liveliness. Users commented variously upon the way these words "just keep coming and coming, like there's someone down there [below the screen] sending them up." As with the subject prompts, which users made sense of by perhaps even granting a kind of sentience to the machine, the active nature of the reading process was perceived to be not only more participatory than a paper letter, but also more "real." This heightened level of participation prompted one of the younger users at the community to comment upon the way in which "that helps me hear what the other person was trying to say. You know? As I move down with the arrow key it's as if the words are coming to me then and there. I don't see them until it's time to hear—kind of like when you got him right there with you. I guess that just seems more real to me."

Up to this point, we have only discussed the act of reading an e-mail message. Even though there is a potential here to study changes in the way individuals engage the machine in terms of information uptake and the manner by which subsequent knowledge representations of senders and the world they relay develop, the act of reading will not be considered any further at this time. There is adequate evidence supporting the notion that the way persons acquire and process information plays a central role in how they construct the reality around them (c.f. Eisenstein, 1979; Greenfield, 1993; McClamrock, 1995; Meyrowitz, 1985; Sigman, 1987; Turkle, 1984, 1995, Alexander, 2005).

It is clear that the introduction of e-mail altered the manner by which some of my Mohawk friends processed information. The question is, in relation to the well-ensconced e-mail user, is it a difference that makes a difference? Not surprisingly, the difference was most obvious when a user generates a fresh e-mail message. A number of very interesting ways of thinking about the process of e-mail arose for them. At the time of this writing, many of the descriptions my research participants made held great meaning and import for them. They describe ways in which thoughts can be corrupted, or made untrue. They tell of pounding temples and thick-heads. They suggest what it feels like to be a Traditional Mohawk sitting in front of a computer screen.

Unlike standard written/typed ground letters via "snail mail," the electronic message typically arrives at the "door" of the recipient just a few seconds after one's finger hits the *Return* key. In effect, the message is transmitted instantaneously. From a temporal standpoint, then, this kind of communication can begin to approximate the causal efficacy of real-time talk, since the transmission can be composed quickly and received essentially without interruption. To quote one of the users, "We can get things done at great distances as if we were there in the flesh." As much as this may sound like any "naive" or first-time user, this kind of talk ends up forming part of an underlying narrative that stakes out what it means to be a Mohawk dealing with some distinctly un-Mohawk things regularly encountered in the world.

With this knowledge comes an awareness that thoughts, words, and information can have an almost immediate impact on the hearer/reader/ receiver. It is this knowledge that prompted most of the users at the community who used e-mail to, in effect, artificially heighten their sense of self as well as their awareness of themselves as communicators, not to mention their awareness of themselves as traditional Mohawk. The term "artificial" is employed here since many reported this kind of thinking to be unnatural and, at times, even uncomfortable. The negative connotations find warrant when Will, a young man who used e-mail only periodically at the community, explains:

> I see what it is there that I'm writing and I often get a little outside of myself. When this happens, you know it's really a kind of scheming. This is a powerful new thing. We need to watch out because it can change minds. Just like if the other person isn't thinking it can really change their mind, too.

Will's emphasis on the mind-changing effects of the technology recalls a time three hundred years ago when indigenous people of this continent were still relatively unaffected by European influences. I recounted the story of "The Mind Changers" above. It was originally relayed to me by a friend of the community, and Tom filled in the details of the story later. I was particularly fascinated by a parallel between one of these early mind-changers and e-mail. Like the violin's alleged tendency to perturb the sacred rhythms of life, e-mail also seemed to be altering something in these people that they held sacred. It was something several of my informants described as an essential Mohawk way of relating to the world. Follow-up conversations with Will revealed that what he implied when he said "if the other person isn't thinking" is that one needs to be careful what one reads on the computer screen. At least three Mohawk I spoke with about e-mail at the time were of the opinion that, where computers are concerned (i.e., the Internet), "there is a lot of untruth in [or rather *out?*] there." As we will discover, there is a sense in which these users are also wary of the incoming messages their e-mail server offers up. We can get at some of what informs this perspective in the way one part-time (summer) resident explains this phenomenon:

> Here's the problem ok? When you're sittin' in front of that complicated machine you have to think 'what's it all *for?*' Well, I know what it's for, it's for the guy sittin' there usin' it, that's what. This thing [the screen] is meant to help you think. That's it. But the problem, you know, with all this is that I begin to think a lot about myself right there. It [the screen in particular] helps too much. I begin thinkin' about myself *as* the guy thinkin.' I'll bet in the end that's not too good, 'cause when you start thinkin' about yourself like that you just end up thinkin' too much. Then you think what the other guy might be thinkin' too and that's when things can really get bad. It can go wrong. You can't do that. The words get sort of twisted you know? So I just try to write my letter [e-mail] all at once, without too much stoppin' and thinkin' and stoppin' and thinkin.' Now, I know you're supposed to go back and see what you said or wait a while before writin' some more but something's wrong with that. But I think that's just the way I see it you know?

The qualification at the end of this statement became something of a patterned response at the community. I think it was a response to the conventional uses of the medium that they became more aware of over time. Indeed, most felt the need to justify their particular habits while online as representatives of Mohawk society and culture. With respect to the ways the medium might alter their perceptions, this possibility points to an interesting twist. Initial interviews concerning the linear and sequential aspects of the medium suggest users were displeased with the way the processor constrains an otherwise free-flowing associative habit of thought. Much to my surprise, this ended up not quite being the case. As Will himself illustrates above, the individuals I talked with became more comfortable with what I eventually described as the *real-time potential* of e-mail. The majority opted to use the medium as a continuous or "quasi-real-time" messaging system.

They chose, in effect, to ignore what many e-mail users in the mainstream value highly. It is, according to many of my colleagues and students, perhaps the most convenient if not strategic feature of any e-mail system: the ability to suspend a message being composed in midstream. Whether or not one buys into the idea of some fundamental difference between this group of e-mail users and others, there is little doubt that they *saw* things differently. In fact, there emerged two distinct perceptions of (and reactions to) the feature of e-mail initially described above as the linear/sequential demand of the medium. Whereas the leader of the community continued to describe and often demonstrate a very direct, real-time interface with the machine (an interface which he reported to be a consequence of his being unable to separate his thoughts from his speech), other users on site accounted for and approved of this perceived linearity and relatively direct sequencing in another way.

The several younger members who used the technology described a kind of pressure to maintain pace with the machine in an effort to avoid the uncomfortable sense of self-consciousness they feel wash over them as the message becomes more thought-out, contrived, or, as one man put it, "a kind of scheming." But, while users report a difference in perceptions with respect to the same formal feature of the technology, most were nonetheless waving their bands in the direction of something which sounded to me like a kind of *oral emulation*. It would seem that users attempted to maintain as best they can while using e-mail the kind of free thinking or open-flow cognitive processes that takes place during their real-time face-to-face talk which typifies communication in and around the community.

I have already mentioned the potentially non-hierarchical or decision-tree-free kind of thinking that may characterize Traditional Mohawk experience. This should not be confused with the free thinking just described. Admittedly, my previous interpretation of many of my research participants' reports was partly erroneous. I formulated that (albeit tentative) hypothesis after a cursory set of interviews with about a half dozen full-time inhabitants of the community fairly early in my investigations there—when these folks have only been using e-mail "regularly" for a few months. The later descriptions lend nothing to any kind of non-hierarchical format. Then again, as this is not a cognitive-psychological study, I will be unable to flesh-out that question with the data I have at my disposal. I am in a position, however, to consider informants' perceptions of themselves. These perceptions seem to describe a wellspring of thoughts and feelings bubbling up from some essential "within" that needs to be let out with as little interruption as possible. "Otherwise," as Will explains, "we tend to present ourselves as something different, something not *us*." What this sounded like to me was simply the awareness that when one does stop and take too much notice of one's message-in-progress, there is a tendency to "see and hear" oneself in the screen. In other words, the visual bias of the technology was regularly cited as *bothersome*.

More to the point, it was as something *not* Mohawk. This kind of forced self-reflexivity was described by most to be incommensurate with their typical way of feeling and communicating. "When you're having a talk, you can't sit there and ask them to wait for you to figure out the best things to say. You gotta come out with it right then and there. So I guess that's what I try to do on the e-mail," explains CJ, an iron worker in his late twenties from the reservation who visited the community during the summer months and was just beginning to experiment with the medium. CJ uses e-mail almost exclusively to maintain correspondence with an ex-schoolmate from the technical college where he received his associate's degree in welding and metalwork. Will, the resident potter, graphic design artist and painter, fills out the notion of "scheming" and, in particular, what it means not to scheme:

> This is not just with computers though . . . when I do my art too. I think this is similar to the way I've always looked at other things—magazines, outside, nature, colors, and then shapes, and I don't remember thinking about anything in particular, it's just getting things down. I know inside of my head those ideas are coming out, but I can't really focus real good on those ideas. Once in a while I'll do a sketch. Just like on the computer. Sometimes I admit I do take a day or two to write a letter. Or I'll paint a sketch. Not very often I'll do that, I'm usually just real free . . .

Other users described something *similar*. CJ talked about "[n]ot spending too much time. You know, thinking and thinking. Like over-analyzing. Because some of the others have said there's something to that with the way Tom talks. I don't watch him much, but I know he don't write speeches ahead of time because he wants them to come off clean and true. He wouldn't even think of it. I think that's similar." Will extends these ideas:

> Yeah, because the more you're free, and immediate, the more you are yourself. Your true self is going to come out. But sometimes with the computer you start schemin' and if you're planning and schemin' then there's blocks here and there. You get that with e-mail a lot. Yeah, actually I come up to a lot of blocks . . . mental blocks when I'm trying to scheme ideas or images. But when I just go freely, it's real fun, it's not exhausting, like scheming is . . . the words aren't the same. Especially when you see them there on the screen. You gotta do what the words say. Does that make sense? With pots it's different. And that's why I like pottery. The medium gives to that. It doesn't change you, change your mind like that [gesturing to his Macintosh PC].

Reflecting upon his experience with art (pottery in particular) in essential juxtaposition to the rigidity of e-mail, Will claims to be wary, and often weary of the latter. Nonetheless, he acknowledges the utility of e-mail and a number of other computer applications, including word processing and an advanced graphic design program he uses three or four times a week. Yet, where the utility of these technologies did seem to attract many of these individuals, they also seemed to recognize what that quite intimate level of interaction with the machine might do to them. We see now how this perceived feature of linearity and direct sequence has led to some very interesting and creative responses by these users. Despite the lack of fading, recursivity, and repetition within even relatively short messages (10 to 15 lines), their communications appear cumbersome and at odds with the format of the message from what might be termed a more "mainstream" perspective. Many messages might even be mistaken for one half of a telephone, or face-to-face exchange. At one sitting, Tom was typing a return message to a correspondent from the reservation, when the buzzer for the downstairs front door began sounding at regular intervals caused him to terminate the session:

>Hello there Wes,
>
>Just a warning to you I am not yet too good at
>this computer. The book is almost complete except
>for a small article yet to be written by Watte
>about the Six Nations Mohawks who were the last to
>leave the valley. The book was not meant to be a
>book but was supposed to be a pamphlet. When we
>began gathering the articles and information the
>pamphlet just wanted to be a book instead. The book
>will belong to the community. The proceeds will
>help the people here to continue our village.
>It will be one of the projects that will help
>promote our self-sufficiency. I have to go down
>to the craft store Wes someone is buzzing. I'll
>talk more with you soon.
>Onen, Tom

What's interesting here is the way Tom pulls ongoing events into his message. There was a similar "real-time" feel hovering about many e-mailings at the community. Despite the other's advice to even leave e-mail messages aside for a day or two, Tom wants to get his message off before going on to other things. Consider this pointer from a friend living several hours north of the community:

>SekonTom
>I have to be honest. When I first wrote I expected
>to find someone else at another location handling the
>communication for you. I was delightfully surprised
>to see you personally responding back. As they say, it
>warmed the cockles of the heart to see your personal
>touch on the message. It's good to see you are on the
>net. It is a lot cheaper and quicker communication than
>traveling. Of course, less taxing on you as well. It
>not as quick as the telephone, but its much cheaper.
>The extra thing I like about e-mail is you have the
>time to prepare your thoughts on the computer before
>sending your message. Sometimes, I leave an e-mail in
>draft form for days before sending it. That way you
>avoid sending a 'flamer.'

Tom had no idea what the term *flamer* implied. I had to explain the concepts of "flame" and "flaming" to him. But, even after elucidating why people usually avoid sending those types of verbal affronts Tom still saw no need to prepare his thoughts in this manner. In other instances he received messages approximating what has more recently been called the "telegraphic e-mail style" in some of the literature. The example below illustrates:

>Hey there Tom
>Good to get your message. *Yeah,* the storm was real bad alright.
>We'll be heading up there soon. Say hello to everybody then.
>onen, Jean

Likewise, Tom finds this style to be inauthentic, or untrue. "I see what he [Jean] is trying to do. He's trying to be fast and simple, but I don't think I like that too much. You might as well tell me something I don't already know." The critique should be well taken. With e-mail and virtually all other forms of digitally mediated communication, many users adopt this kind of punctuated, telegraphic style. Be it pressures at work, or just last-minute intentions to get a message off, many of us have taken to what, on the face of it at least, has become a much more impersonal form of communication: to be short, to the point, and "practical." It seems to be standard operating procedure for so many in Digination. But this move toward practicality, according to most of the folks I interviewed, will have to result in a colder, sterner, more sterile interaction. And so here is where comparisons with the native language become fruitful. As Tom expressed to me on numerous occasions, it is not merely the format of e-mail messages that can contribute to this coldness—some of the coldness or harshness is within the English language itself.

For Tom at least, the Mohawk language holds the key to inner feeling. For instance, as the first example he mischievously provided me very early on in our interactions illustrates, a Mohawk term implying sexual arousal (i.e., "having the hots," or "feeling horny") communicates to the hearer, "I am hungry for you in a sexual way such that if I don't have you now I will surely die." In comparison to this, English is a stern functional language according to Tom. The story about how he used to fear death as a little boy when he heard it talked about through the Western/English lens was relayed to me through watering eyes. You just say "when people die we bury them." For Tom, "all one feels at this is the death itself." But, he explains how life (and death) really is richer and warmer than that. In the original language, we are told that "we must prepare our loved one to be wrapped in the green, leafy field blanket of our earth mother."

So Tom resists a kind of distillation embedded in the English language itself. But at the same time he resists the logical sequence of the written from, as the ability to suspend messages by simply typing as he talks. Indeed, the aim in all of this seems to remain the same—namely, to liven up the interaction. However, many mainstreamers (or what might be termed "assimilated users") actually resist the linearity in just the way Tom's friend suggests by composing messages over a period of days. This is a practice radically at odds to what Tom prefers. He sees e-mail as serious and important, but in a different way than his more assimilated friends. With this discussion in place, we are now in a position to draw out a tentative taxonomy of e-mail styles.

Several strategies arise when we view a cross-section of e-mail use on and off the settlement. While less indicative of ostensive differences in use at the community and in the mainstream, the general opinion among the inhabitants who use the medium seems to be that this is the way it is supposed to be. For them, the first and third types—*telegraphic* and *composed*—best characterize communication elsewhere, whereas the *Oral/real-time* style describes things at the settlement fairly well:

Telegraphic: This is a short, punctuated style that complements the speed of digital communication as well as the surface linearity of the processor. In the mainstream, this seems most prevalent on college campuses, in businesses, on LANs (i.e., local area networks) or inter-office systems and, somewhat counterintuitively perhaps, often between the closest of friends. Users do not have to abide by the linearity of the processor, but we are seeing this as a strategy of communicative efficiency with e-mail and especially text messaging today.

Oral/Real-Time: Tom exemplifies this style. While he abides by the linearity and continuous sequence of the processor (uses e-mail in real-time), he resists the logical if-then feature of writing in general as well as the telegraphic tendency to encapsulate thoughts in as small a space as possible. His style of e-mailing, however, is noticeably different from that of some of the more assimilated *Indian* users encountered on and off the settlement.

Suspended/Composed: Messages are composed over a period of time—sometimes even several days. A user might adopt this style/strategy to enhance his/her image within the medium, and/or to avoid "flaming" and a variety of other *tele-faux pas.*

So I was coming up with definitions and categories and putting data into comfortable slots, but still wondering how I might account for some of these Mohawk perceptions of what was already (in 1997) becoming a fairly mundane communication technology. While they seemed to be less isolated from the mainstream world than the several dozen Amish families living atop the escarpment about five miles to the northwest, there was still a very small amount of communication media in use during a typical day at the Mohawk community. This is partly due to the daily activities that constituted life there at the time, such as farming, planting, animal care, and general maintenance of the fields and other (primarily physical) work requiring a lot of mobility. Only when individuals were working indoors, or in an area where they remained within the same several square meters (as when the horses were being tended, or a tractor serviced) did I find them using a radio or cd/tape player. These other electronic media served much more of an ambient function in such instances, and this appeared to be not all that different from such media use in the mainstream American contents.

But, with the computer and e-mail specifically the local impression was that there were some fairly specific dynamics at work. For instance, the linear format of the written language itself is seen as having one effect. Most users were of the opinion that this format forces a kind of rigid structure onto their thinking. In keeping with their much longer media history, these Mohawk expressed their concern that written (and in this case typed) language is a powerful force that continues to exert pressures on the more natural ways they think, talk, and communicate. For them, actually seeing their words brings with it a felt need to make sense and be logical that goes beyond that experienced when talking. But, we need to understand that "logical" in this case does not necessarily implying greater *sense.* It merely implies more accordance with the strictures of the e-mail interface, akin perhaps to the logical form of the proposition or syllogism. The Mohawk I talked with about this seemed to be keenly aware of the power of the medium to "change minds."

These descriptions are telling and clearly served an ethnomethodological function for these nascent community builders. Users noticed that their way of seeing and being in the world is altered at a very active level when mediation includes any kind of literal, logical, highly structured representation. Their personal accounts include descriptions of uncomfortability and general unease. It is as if the world is being parsed out or chunked in ways unnatural or unreal. There is an underlying rhetoric of artificiality, indeed a lack of authenticity in users' accounts of their interaction with e-mail.

Participants acknowledged their distaste for the feeling of heightened self-awareness. In light of this they did their best to avoid or suppress that feeling by typing in real-time and even leaving the dialogue box small so as to increase the fading potential or ephemerality of their compositions. That being so, it is reasonable to think that there must then be an active level of self-reflexivity at work if they are in fact able to resist some of the mechanisms described above. Indeed, there seemed to be a difference in "representational competence" (Greenfield, 1993) between these symbol systems. Greenfield proffers a compelling description of the way particular channels of communication may alter subsequent "pictures" being drawn. Given this, I had to admit that there was a kind of media determinism at work here.

However, I will finally urge that we dub this a *soft* or *weak* determinism since users exploit the opportunity to resist these formative properties of the medium. While they claimed their interactions with the technology could be difficult and often taxing on them, they were able to exert their own agency while interacting with the technology. The meaning of the interaction was not decided, as it were, by the technology or medium itself. As far as these Mohawk were concerned, gaining representational competence, or fluency or "literacy" in e-mail seems to lead one to think about ones thinking in an unprecedented fashion. The experience is a heightened sense of one's ethnic and/or cultural position vis-a-vis the technology. I think these descriptions provide fair warrant for at least a perceived difference between Mohawk and mainstream users. However, these perceptions were not without consequence. They led to a functional difference in the way my research participants communicated when using e-mail.

It is for this reason that I became particularly interested in the way these individuals not only used e-mail, but also *talked* about their interaction with the technology. And for other technology users elsewhere, the piece of

Digination that is about *dignation* is a very important piece. I think e-mail, being in so many respects incongruent to Traditional Mohawk things, provided them with an experiential space or venue around which a great deal of cultural talk was generated (c.f. Carbaugh, 1996). The general perception at the community was that the technology *works* on these individuals in a manner they would rather avoid. Nonetheless, many chose to use it. Functionally, the screen in front of the user, along with the knowledge of speed or instantaneous transmission, exerts the weight of hearing, seeing, and otherwise positioning oneself culturally as a communicator in those ways described above.

NOTES

1. There are a number of reasons why the term *reconstruct* is used here. For one, there is the problem of recovering the lost practices of an oral tradition. The oral means of cultural preservation are naturally subject to continuous revision and updating. This is why, in some respects, the *re-construction* is a *re-invention*. In addition to this, the term echoes one placard in front of the property at which this group of Mohawk resides that describes the community as having recently been "re-established."

2. While "Email," "e-mail," "eMail," and "email" have all been conventionally used to refer to the technology in just the last decade (both in academic circles and the popular press), "e-mail" will be used throughout this book.

3. The original Carlisle Indian Boarding School operated in Pennsylvania throughout the 1800s. As of late, however, the term *Carlisle School* has come to represent, for many Iroquois, every institution that sought (or less explicitly continues to seek) to systematically eradicate any vestige of Native thought and culture in its student body. These institutions were actually very common throughout the United States and Canada into the twentieth century. Conversations with Tom suggest that he believes such institutions still exist under more subtle guises.

4. One such proverb: *"Learn a new language and get a new soul."*–Czech, *anon.*

5. The term "culture" as employed here is analogous to an "interpretive community" (as articulated by Lum, 1996; and Radway, 1996). According to these authors, an interpretive community is a group of users who are defined primarily by their collective interpretation of the medium or technology being utilized.

6. The use of the term *a priori* here refers to the social scientific meaning that implies knowledge or awareness that is *tacit, assumed,* or *given* (as opposed to the standard philosophical usage which implies knowledge that precedes any direct experience).

Chapter 4

Blogs

The New News Medium

"A point of view can be a dangerous luxury when substituted for insight and understanding."

—Marshall McLuhan

In the previous chapter I detailed the core observations resulting from a micro-analysis of the perceptions of individual technology users. Aside from a review of some of the pertinent literature, I tried to keep the theory in the background so that my research participants' perspectives had a clearer voice. In this chapter, I hone the theoretical lenses a bit and proceed with a more formal approach so we can grapple with some of the conceptual confusion surrounding blogs, or web logs, a popular and still proliferating media form today. Employing the medium theoretic approach already described, and a symbolic interactionist perspective that will help draw out some of the relationships between individual and group identity; here I consider several key structural features of blogs and discuss some of the personal, social, and political significances of news blogs and blogging. To begin, consider this little piece of media history.

In 1690 Benjamin Harris published the first issue of *Publick Occurrences Both Forreign and Domestick* in Boston. Harris' project is generally recognized to be the nation's first newspaper. Each printing was just four pages long, with *page 4* intentionally left blank. With a literacy rate very close to one hundred percent among the Puritan population of the time, these papers were typically passed along a string of readers (and contributors) during the day. The blank page was provided to allow readers to write in comments or news updates before passing it on to the next citizen that happened along. If this practice did not continue beyond the first edition, it was still a precursor to the twenty-first century notion of "interactive journalism" in general and the news blog in particular. So we know that the idea of the citizen journalist has been ruminating in the public mind for some time.

Today, news Blogs (web logs dedicated to the dissemination of news) are becoming the default political news source for a growing number of well-educated and ostensibly well-informed segments of the population. Bloggers and blog advocates suggest that blogs, online lists, and their analogs offer something different and potentially unique to the twenty-first century citizen. At their best, blogs represent a new form of open-source/open access partisan press that promises to bring McLuhan's global village one step closer to fulfillment: a vibrant, interactive polity resembling Jurgen Habermas' Ideal Speech Situation where differing minds engage one another without fear of reprisal. Blogs have certainly been instrumental in helping oppressed or disenfranchised populations get word out to the world. The blogging waves in Iraq and Iran over the past several years is testament to the ability blogs and bloggers have to flatten power differentials and help the world witness myriad injustices and atrocities. The same applies to blogs (and their bloggers) associated with the *Arab Spring*. However, at their worst, blogs also represent the latest form of mass-mediated triviality and celebrity spectacle, with the potential to create and sustain insulated enclaves of intolerance predicated on little more than personal illusion, rumor, obfuscation, disinformation, and politically motivated innuendo.

There's no doubt, blogs are funny things. They blur the distinction between what has traditionally been conceived as public and private information, between the individual and the group, and between fact and fiction. Then again, for every detractor who makes note of the potentially diversionary role played by the Internet and blogs, there is a champion who posits that "blogs act like a lens, focusing attention on an issue until it catches fire," or that blogs "can also break stories . . . on April 21, [2004] a thirty-four-year-old blogger and writer from Arizona named Russ Kick posted photographs of coffins containing the bodies of soldiers killed in Iraq and Afghanistan and of Columbia astronauts" (Grossman and Hamilton, 2004; brackets added). It is likely, in fact, that many, even all, of these positive and negative observations regarding blogs have credence. Search engines like Google and Yahoo are still figuring out just what to do with these new news manifestations, so a more global assessment regarding the role of blogs must be deferred for the time being.

One thing is certain: the battle over blogs is in full swing. So what are blogs and how should they be defined? Who is participating and why? And what role are blogs playing today (what role can they play) in the news making of tomorrow? How might blogs play a part in the political process? Or, how can blogs be exploited in such a way so as to prompt substantive reform in both journalism and politics? To get an initial handle on these questions we'll consider several leading scholars' thoughts regarding how a social and political agent—a citizen and civic participant—*comes to be* online.

With a special focus on political news[1] we know that blogs open up the potential for a diverse set of perspectives that can broaden the public's knowledge base and essentially flatten the hierarchical feel of traditional, mainstream news outlets. At the same time, however, I'll caution that a blog's underlying structure, and its editor (a kind of ever-present virtual shopkeeper) provide a certain trajectory to the discourse that unfolds. Blogs have continued to morph since they were introduced just about a decade ago, and while there is not a strict editorial protocol on most blogs, there is a tacit form of control that may be at odds with the ostensibly open structure of the *digital commons*, that ideal but entirely over-used characterization of what we know today as the World Wide Web.

It was Aristotle who first noted nearly 2500 years ago that all communication is persuasive communication. I think we can sharpen this characterization today in our new context of Digination and say that all communication is political communication. Indeed, I'll demonstrate why communication on blogs in particular finally brings Aristotle's statement into its own. If there are plenty of folks in and out of academe studying political communication who think that blogs represent an evolutionary step toward truly democratic political processes by freeing the public from the commercial constraints of corporate media, a nagging concern lingers with respect to the way blogs organize contributor's thoughts, and structure and limit the interplay of ideas. With significant help from two seminal communications theorists, I'll proceed now in offering one informed perspective on the personal, social, and political significance of blogs. We begin with one quite attentive blogger's take on things.

> As bloggers, we update our sites frequently on the content that matters to us. Depending on the blogger, the content varies. But because it's a weblog, formatted reverse-chronologically and time-stamped, a reader can expect it will be updated regularly. By placing our email addresses on our sites, or including features to allow readers to comment directly on a specific post, we allow our readers to join the conversation. Emails are often rapidly incorporated back into the site's content, creating a nearly real-time communication channel between the blog's primary author (its creator) and its secondary authors (the readers who email and comment). (Hourihan, 2002).

Several observations made by blogger Hourihan are worth mentioning. The first is that bloggers tend, almost by definition, to comment only on news and information that they personally deem worthy of public note. The second observation is that the near-real-time open accessibility of many blogs (with the comment option at the bottom of the page) makes them a collaborative form of news and information quite distinct from a newspaper's letter to the editor or op-ed piece. However, the apparently "open" format of the blog is at the same time a quite rigidly organized communicative form. I never created my own blog before publishing the article (in 2005) upon which the present chapter is based. But I did accomplish a lot of lurking and snooping around on blogs of various sorts.

I wanted to make news blogs the focus of this chapter at this point in the book because I think it is in the realm of news where blogs will make their most significant and enduring contribution to society. It is with this observation in mind that I formulated the following questions: Do news blogs really host a new form of collective journalism? Are they hotbeds of social activism or just a new form of *slacktivism,* as a recent contributor to the *Economist* magazine opines? Are they just thinly veiled catalogues of self-referential meta-commentary concerning the host blogger? Blogs just seem so multi-functional. According to Jill Walker Rettberg in her book *Blogging* (2008) it would appear that blogs can do it all, and well too. It seems blogs can be home to all of the above—and more. As a renowned blogger in her own right, I suspect that Rettberg is subject to the anthropologist's standard critique that *whoever discovered water, it probably wasn't the fish.*

The point, of course, is that it's notoriously difficult to understand the nature of an environment from within—I mean, as an *insider.* While Rettberg is also an accomplish communication researcher, I'm going to argue that her perspective (akin in many ways to Nicholas Negroponte's take in *Being Digital,* is colored by an interest in making that way of being good and right and true. So I'll temper some of Rettberg's claims regarding the emancipatory potential of blogs here and there by pointing to several aspects of blogs that she chooses to ignore. As far as I can tell, Rettberg suggests we dismiss the technological determinism thesis in its entirety. I'll submit, if I'm reading her correctly, that in doing so she throws the baby out with the bath water. Rettberg glosses formative or causal significance of the design features and properties of blogs. Indeed, she seems to think about all technology in this way. As Rettberg puts it, "[a]lthough it is clear that technology does affect the ways in which we live, technology does not appear out of a void, and is itself shaped by cultural developments" (2008, p. 52). So there's this balancing act going on and humans tend to figure things out in their favor after a little trial and error. Right around this point in her book Rittberg very diplomatically dismisses the determinists claim that technical systems can often exert unequal force in the human-technology equation. Her argument hinges on the following assumption that has stymied our efforts as a species to understand our relationships with the tools we build and disseminate out into the world: since we design and build technologies we can and eventually do control them, or at least renegotiate their design and their impact on society in the process of use so as to settle on a beneficial outcome. Again, I think this line of thinking gives short shrift to the systemic or ecological nature of technology in general. It's the idea that once an artifact is introduced into a system, everything changes. History shows us this reality in spades. Consider the outcomes and unintended consequences of innovations ranging from the firearm, to the automobile, to the cell phone to the credit default swap. Nope, I'd say we have yet to *get a handle* on any of these. Not a one. In my eyes this theme of technics-out-of-control has become a law of life in Digination.

THE HULLABALOO ABOUT BLOGS

There are millions of blogs on the Internet. In 2005, it was reported (on several blogs devoted to statistics on blogs) that a new blog was created in some corner of the Internet every seven seconds. In 2009, *Technorati.com's* "State of the Blogosphere" suggests very similar numbers. While this equates to more than 12,000 new blogs being added to the Internet each day at the time of this writing (an almost inconceivable number), I do my best in what follows to focus the discussion on those blogs that purport in some way to be legitimate political news blogs. Several examples of this kind of information node that I lurked on include: the *Huffington Post, Online blog,* and *Culture vulture* (at *blogs.guardian.co.uk*); *blogospherenews.com* ("Blog News from Bloggers, for Bloggers, delivering the latest news from the Blogosphere"); *Blognews* (at *topix.net*), *www.breakingnewsblog.com, Slate.com*; *News Blog* (at *OnlineJournalism.com*), *beldar.blogs.com* ("the online journal of a crusty, long-winder trial lawyer, bemused observer of politics, and internet dilettante"), the *Huffington Post, Townhall.com* and *Poynter Online. Poynter* had an interesting lead article that posted on June 16, 2005, entitled "The Blog-Only News Diet" that is periodically referenced below in an effort to illustrate some key features of Digination.

Before focusing our theoretical lenses, however, we should consider the nature of the debate over the online communities so many of us participate in these days—everything from using e-mail as a sort of mini-listserve, to sending *e-vites* or *tweets,* to hosting your own website, *MySpace* or *Facebook* page, to participating in a bona fide listserve or blog all fit the bill. Given the increasingly ubiquitous role the Internet now plays in so many of our lives, some sociologists and communication theorists have argued that our interaction online is destroying real-world (or "off-line") community groups and voluntary associations, and has the potential to divert the citizenry away from traditional political processes that are necessary for a democratic society to succeed and thrive (Bowers, 2000; Carpini, 1996; Postman, 1992; Putnam, 1994, 2000; Rash, 1997; and Turkle, 2009).

Some of these commentators make explicit reference to blogs; some dedicate their analysis to political activity generally, but most provide a more general critique of the now very mundane act of going online and engaging in decentralized, disembodied, temporally-displaced, "social" interaction. They point, in other words, to processes inherent to the idea of Digination. And a great deal of empirical research focusing on the Internet as an emerging public information sphere has been undertaken over the past decade. For instance, Deuze (2003) considers some of the consequences of different types of online news media. He highlights three features of Internet-based news that speak directly to blogs: *hypertextuality*, *interactivity*, and *mutlimediality*. Deuze wonders if we can even group online journalism alongside traditional print-, radio- and television-based journalism. Due primarily to hypertextuality and interactivity, I tend to agree with Deuze's tentative conclusion that we cannot. Indeed, from a media ecological perspective, blogs are as different from traditional (text-based) journalism, as print is from television news. In a related study, Selwyn (2004) reconsiders some of the political and popular understandings of the digital divide—the fuzzy line that presumably separates the "information haves," from the "have-nots." Selwyn's analysis is useful because it prompts several more nuanced definitions of information access and interactivity, with the latter concept considered more appropriately along a continuum. Subsequently, simply having access ends up being a spurious, and therefore inadequate measure when it comes to predicting an individual's actual knowledge base and real potential for informed and substantive political participation.

Taking a close look at the nature of discourse online, Salter (2004) observes that while "it is true that the structure of the Internet may well facilitate certain 'democratic' forms of use, this is not a necessary fact" (p.1). Salter's analysis is valuable because it raises the possibility that the way the medium is used and the way it is talked about might have more to say about its current and future structure than the way it is formally designed and imple-mented. This is a key observation that gains additional credence in the present study. Extending the general argument that Internet-based communication is becoming increasingly consequential to "real life," Shefrin's (2004) analysis of fandom and the culture of media entertainment highlights the relationship between new modes of authorship, production, marketing, and consumption on film and celebrity Internet fan sites. Employing Bourdieu's theory of cul-tural production as a primary analytic lens, Shefrin suggests that happenings online may be influencing the offline world much more significantly than many commentators are willing to admit. Bennett (2003) offers an exhaustive analysis of the various strengths and vulnerabilities of online politics. Blogs are considered by Bennett to be a kind of "middle-media," often forming a bridge between overt political action sites, forums and newslists (what the author calls "micro media"), and mainstream news outlets like CNN, MSNBC, and the New York Times (i.e., "macromedia"). Bennett's work helps paint a clearer picture of the blog's role in the emerging political processes of the twenty-first century and should prompt a renewed look at Gitlin's (1980) analysis of organizational dissolution in the face of mainstream mass media cooptation. Through several case studies Bennett reveals, most notably, how the rapid expansion of international activist networks depends upon relatively weak ideological ties for their sustenance, and so is reminiscent of Granovetter's (1973) observations concerning the strength of weak social ties as a binding force for organizational longevity. While focusing on the present and near future of political issue campaigns (micro media-based political and social movements), Bennett's analysis is useful to the present study of news blogs because it demonstrates how weak social ties and loose identity affiliations—the same qualities that make most forms of *virtual politics* durable through time—also makes them potentially vulnerable to problems of control and emergent group-think. Sharpening many of these same observations regarding the strengths and weaknesses of virtual civic spaces, Myles (2004) undertakes an original ethnographic research project which looks into questions of agency, ownership, and power on several small-scale civic (city, town, and county) Internet sites in the UK. Myles interrogates a broad array of social policy issues including the commodification of local electronic civic networks, the influence of "cultural intermediaries," and the impact of an entrepreneurial ethos in the development of community networks. Myles' work bolsters the present study since, on news blogs, the roles of near-omnipresent editors, as well as sporadic contributors, remain hotly contested issues.

Rettberg (2008) does her best to parse out blogs according to a number of variables like intent, content, and authorship. She cites significant differences between corporate blogs, citizen journalist blogs, blogs as narration, blogs as self-exploration, advertising blogs, free blogs, pay-per blogs, etc. As another manifestation of the neutral theory of technology, I think this misses the forest for the trees. Like any communication medium, blogs embody certain properties that exert causal force which both alter their content and the people who use them. All blogs are narrative in form. They are also all "corporate" and advertising-based in the broad sense of the term because they are all self-referential or promotional in one respect or another due to the way the author is bound up in the linguistic stream that is available for all who are interested to see. Indeed, this gets us back to Aristotle's dictum about communication. It is all inherently persuasive in nature. On blogs it just kicks in to high gear.

In addition to the mounting number of academic works on the subject, the general topic of bottom-up online news and information resources has produced steady commentary about blogs in the popular press over the past five or six years. Beyond this, there is a near constant conversation taking place on mainstream Internet news sites about blogs, not to mention the almost incessant self-referential meta-commentary about blogs on blogs concerning their journalistic integrity and character, their news function, and by extension, their social, economic, and political utility as information repositories. However such observations and questions rest on several more fundamental issues yet to be seriously considered in the literature. In order to understand how an individual's or group's character can be developed and/or dissolved on a blog, we need to engage in a more sustained analysis of the roles played simultaneously by the content and form of this new manifestation.

SOCIAL STRUCTURING ON BLOGS

There is a simple question we can ask about blogs. It is also an ancient question reapplied: *do birds of a feather flock together or do opposites attract?* If both, then why and how, in what contexts and under what conditions? It is generally understood in most of the ethnographic and survey research to date on the subject that online fora like blogs, discussion boards and listserves tend to engender more homogenous than heterogeneous groupings and social affiliations. Such a finding seems intuitive enough since it extends the homophilic tendency in humans (and essentially all species on this planet). However, what role might the formal structure of the system, itself, have to do with such a tendency? Some work in this area is getting accomplished, and some preliminary answers have been offered. For instance, stemming primarily from the perspective of the user, Norris (2004) notes that on the Internet "[c]ommitments to any particular online group can often be shallow and transient when another group is but a mouse click away" (p.33). Online communities that do not have a physical counterpart are usually low social and psychological cost, "easy entry, easy exit," groups. "To avoid cognitive dissonance," says Norris, it is often easier to simply "exit than to try to work through any messy bargaining and conflictual disagreements within the group" (ibid).

I think Norris' critique remains well-grounded and continues to be illustrated in the way blogs are used by the public. That is to say, blogs do trend toward homogenous groupings. It almost seems a bland truism to point this out. For this reason, political blogs might be viewed in a positive manner by their advocates as something like Bennett's (2003) "ideologically weak networks," that can reduce the conflicts often associated with diverse players. Political blogs are what they are: explicit partisan sites intended to extend the message of a candidate or interest group (consider *MoveOn.org*, or *Swiftboat Veterans For Truth* during the 2004 U.S. presidential campaign). However, news blogs and online news lists often claim to have no explicit agenda other than disseminating information. But due to certain structural features of blogs the "open objectivity" some of these online fora purport to have may systematically degrade over time into thinly disguised partisan platforms, thereby becoming ideological nodes in a network of what, on the face of it, appear to be open-sourced (i.e., polycentric) and openly accessed political news and information repositories.

Blogs are still too new on the communicative scene (the first surfacing in the late 1990s) to get a clear picture of their wider import at this time. But it is also that blogs are more than just technological innovations—they are complex environmental systems, bona-fide socio-cultural-political phenomena. For this reason we really need to learn more than just the "how to" of blogs—how to build a blog, how to ensure that your blog places high on a search query, how to generate positive cash flow from your blog. There is plenty of this sort of information about blogs out there already. More and more people are learning how to build and use blogs. Bloggers who know how to "game the system" might have the most visible or loudest "voice," but I'm still not convinced we have a decent handle on what this particular manifestation of our Digination entails.

Recall the short article written about the "blog-only news diet" for *PoynterOnline*. *Poynter* regular Steve Outing features Steven Rubel in the piece. Rubel is a public relations executive who did his best to bracket his exposure to news and information about the world. Rubel allowed only blog content through the filter of his conscious awareness for a period of one week. He only read what bloggers wrote. I'm hoping the reader also detects something strange about this little experiment, because there is something very artificial (and methodologically suspect) about Rubel's self-imposed prohibition to utilize any hotlinks contained in the blogs he visited. This is akin to reading a news article and skipping over quotes from people interviewed by the reporter, or perhaps walking to the kitchen and fixing yourself a snack during a television news cast any time the anchor turns to engage in a brief Q&A with a correspondent live at campaign headquarters. Blogs are themselves constituted by the hotlinks they reference. Any effort to constrain blogs to the content residing at their first level is to obliterate the definition, meaning, and function of the blog.

Later in the article both Rubel and Outing critiqued the blogs Rubel visited for not being, essentially, high-tech enough. This is a notable observation as it reflects a common interpretation of the Internet in general:

> There's not a great deal of innovation in terms of storytelling technique. We're still talking about text and a few static images, mostly. (Rubel did see the widely distributed photo last week of President Bush fighting a wind-damaged umbrella, for example.) But when it comes to audio, video, Flash, and interactive content—for the most part, forget it. (Outing, 2005; p.2).

Good story-telling technique, according to Outing, redoubles Rubel's own opinion that the default text-only format is a systematic shortcoming of blogs. I'll disagree with their assessment and suggest that the text-only format may be what gives blogs their special power today. Indeed, Rubel's statement works well in framing the heart of the matter as far as McLuhan, Media Ecology, and the philosophical approach of this book is concerned. It is that the formal properties of the medium help dictate the kinds of information/content that is transmittable by that medium. The formal structure of a medium also does much to frame interpretations and predict the kinds of meanings attributed to and gleaned from that content by those "attending." If so, then the following questions emerge: does limiting the presentation of information on blogs to the textual format actually limit, or even degrade the quality of, the story being told? And does this, by extension, limit the ability one has to attract and maintain audiences? Does "telling a story" mean getting something specific across to the audience/user with regard to the *reality* (the world "out there") being referred to? These are intriguing questions, and they sharpen the *meta-talk* of Outing, Rubel, and countless others working to make the future of blogs a certain kind of future (c.f. Salter, 2004; Rittberg, 2008).

But if blogs rarely seem to be short on political commentary, Outing does cite a decided lack of financial and business content in the blogosphere. He argued in 2005 that a new blog niche can be found under those two subjects. I think his professional advice has yet to be sufficiently taken up—at least not seriously, not in the most effective form. Instead, we are seeing a new kind of blog taking shape across topic areas, and it is a very recognizable sort of beast. Blogs are turning into vlogs (video logs) and other sorts of image-centric narratives. I am certain this is where things are headed not because I possess any special knack at divining technological and business trends—if I did, I might not be writing this book on the paltry retainer that is my professor's salary. If we look at the larger patterns, the texting trend does not detract from what seems to be the "natural" development of any digital information environment: written language in such an environment continues its slow creep toward obsolescence.

Now, one problem according to Outing and Rubel is that business and financial news tends to offer all but the "dedicated insider" notoriously dry material. In other words, it is usually just *words*. Given the call from Outing for "audio, video, *Flash*, and interactive content," the idea seems to be that morphing pie charts, swooshing PowerPoint-like presentations, and video interviews with key spokespersons could make such material more attractive to a restless virtual public increasingly accustomed to high-tech displays for the eye and ear when they venture online. To be sure, headline news stories over the past several years detailing everything from the Enron fallout, the housing bubble, the ensuing systemic economic collapse, and certainly the recent healthcare debate are often flat, laborious missives that get bogged down in the minutiae of the hour. They often include quite impenetrable language surrounding things like hidden ledgering, exchange rates, junk-bond valuations, credit default swaps.

Again, Outing and Rubel's critique is interesting because it runs directly counter to some things Marshall McLuhan said regarding the characteristics of various media. As McLuhan would have it, the textual mode can serve us well in such instances with its built-in sequential logic, and cause-effect/ subject-predicate grammatical forms. Instead of attracting audiences with audio-visual magic shows, it could just be a matter of including more pithy and apt expressions like: *"Ken Lay is the kind of guy who could sell junk bonds to hungry dogs off a meat truck"* or *"A good number of mortgage lenders were just stealing from Peter to pay Paul"* in such stories to make the difference between dazzling them with high-tech displays on the one hand, and getting them to think about and understand the issues, including the reality and relevance of corporate irresponsibility in their own lives today. Following McLuhan's lead, the tetrad at the end of this chapter suggests that news blogs in their best guise (that is, with a text-biased or word-based format), retrieve from the past an appreciation for the aesthetics of language, and with it, the wordsmith and intellectual as social commentators worth listening to.

McLuhan made some intriguing observations regarding the power of mnemonic expressions and their constitutive function in oral cultures. He also had some compelling insights into their vestiges found in our own waning print culture (see also Ong, 1972). All of these points McLuhan made are instructive. Even in the context of political news there is something to be said for succinct yet masterful language when used strategically. It is, in this way, that the blog reintroduces us to the aesthetics of language which began to diminish with the telegraph, lost significant artistry and traction on national audiences throughout the golden age of radio, and has been progressively effaced from the civic sphere with the advent and ascendancy of television. Yet for now blogs continue to offer the public (or, in any case, that portion of the public with a nominal Internet connection) one set of venues for this kind of down-to-earth, intelligent, pointed, audience-aware writing.

There are two blog tetrads illustrated in the appendix of this book to suggest how the heuristic can also be focused onto specific levels of analysis for any medium under consideration (in the blog case I described the technological and the socio-political levels). Interesting overlaps and some surprising distinctions always emerge, and it would be interesting to see what the motivated reader comes up with in terms of alternatives to the single tetrads for the remaining chapters where a medium is highlighted—or even alternatives to the two levels of analysis describing blogs. These alternative groupings and clusters are not only useful to think about, they're also pretty fun to create and play around with. I can imagine this sort of exercise becoming standard fare in, say, any high school computing, social studies, or history class. It's sometimes a little hard to get started (and I find that students typically start with what is enhanced by a technology), but once things get underway the tetrad quickly mushrooms . . . or perhaps it blooms, I'm not sure.

Again, the first tetrad I drew up for blogs details the *technological level* of the news blog and suggests a more traditional medium-theoretic reading of the technology. In addition to the enhancement of memory and intellect, and the potential to obsolesce the entire corporate news edifice, this tetrad suggests how the news blog retrieves (if only temporarily) the rational discourse of an earlier time due to the logic, and cause-and-effect descriptions that are intrinsic to the linear, sequential format of the printed (or in this case teletyped) word. However, given the high-tech dictate of the Internet synopsized well enough in the Outing article, we can see how primarily text-based blogs may ultimately reverse into (unintentionally "flip into," or become reconfigured as) wide-band multi-media nodes not at all unlike any mainstream news site currently in operation (c.f. *CNN.com*, *USA Today.com*, etc.).

Blogs, according to McLuhan's standard tetradic recipe, will likely follow the evolutionary pattern of virtually all public-oriented information on the Internet. They will continue to morph and find their form as multimedia, interactive sites—since to be any less (technically) is a recipe for perceived obsolescence. To be sure, we are already seeing this trend such that now if someone (an individual or a collective) has any hope of being taken seriously as a presence on the Internet today they must incorporate images in their pages—many of them, and moving ones preferably. This might perturb aspects of McLuhan's suggestion that old media technologies tend to become the content of new media. Blogs are morphing into Vlogs and words are getting as scarce on many Vlogs. However, there is no recognition in Outing's article of the function of different forms of media, and the different formats of representation various media tend to default toward. Just one mode—the visio-pictorial—is alluded to be the best online.

THE IMPRECISION OF THE IMAGE

The default representational form of political news on blogs continues to move toward the pictographic (moving and/or still). I'm arguing in this chapter—and I will expand upon this argument in subsequent chapters—that some serious problems will persist if we are interested in maintaining a public that is willing and able to participate substantively in the political process with such media forms. And this sort of argument extends far beyond McLuhan's public musings. We seem to use both innate and learned logic and rationality in the language interpretation process.

With new research in the cognitive and neuro-sciences still emerging which supports the general notion that communication media can alter cognitive, mnemonic, and perceptual function, there may be a trend occurring online that speaks to of Abraham Maslow's happy phrase: "It is tempting, if the only tool you have is a hammer, to treat everything as if it were a nail." This is the *law of the instrument.*

While not yet the undisputed default mode on blogs, the image is becoming the standard method of relaying news and information on the Internet. I'll go into more depth with this and the brain science support in another chapter, but for now a couple well-known media examples should suffice. The image of angry Iraqis pummeling the fallen statue of Saddam Hussein in the middle of Firdos Square in Baghdad on April 9, 2003 is one obvious instance of the limits of relaying news via pictures. Virtually all American news conduits selected a series of tightly framed close up shots that told a very particular story about the occupation and toppling of Saddam's regime. It seemed quite obvious from these repeated portrayals in the news that the Iraqi people enthusiastically supported the American military presence. However, many commentators felt that, were it not for subsequent discussion about that event on blogs, the fuller reality of the situation would have been lost in the thick tumult of information with a heavy bias on image-based information being exported daily from Iraq.

Blogs were another way many interested observers became aware of the larger reality beyond the image frame in Iraq. The manner in which the event was staged, the way excited Iraqis were corralled close to the foot of the statue in the square, and the tight camera angles were all facts largely ignored by the mainstream press in the United States. The photo-op of President Bush's Top Gun landing on the deck of the USS Abraham Lincoln in about three weeks after the fallen Hussein statue story is another case study in this regard. That pseudo-event (Boorstin, 1961) requires no elaboration at this point; however, Bennett (2005) provides a cogent analysis of the way in which the "news reality frame," now significantly enabled by imagery, can effectively short circuit the public's ability to make sense of unfolding events.

So if blogs were meant to provide a firmer grounding to the photographer's whim and the decontextualizing tendency of the image, we can argue that there may be an inherent logic embedded in the Internet wherein the most technically sophisticated mode of symbolic transfer (the image) becomes the mode of choice.[2] Thus, we can invoke Maslow's *law of the instrument* and see how, when a news item is hammered and battered online via imagery, the subtlety and nuance that lays beyond and behind the images we see can be lost. No doubt, the meaning—including that of operative statements and deeds made by key political players—can easily get lost in the "1000 words" a well-framed picture has to offer.

But that is perhaps only one future history of blogs. Leaving this particular McLuhan-esque prophecy as it is, we should explore next the political and epistemological functions of the current format of news blogs. For now, many self-described political news blogs still maintain an *image*-to-*word* ratio that easily favors the latter. If the seduction of the image is not, as yet, a serious issue with this type of blog, there are some other formal demand characteristics associated with primarily text-based new formats online. Words have their own problems here.

THE TYRANNY OF TEXT IN THE BLOG THREAD

With the tetrad we can begin to see how the manifestation of ethos and identity on blogs (both for individuals and larger groupings) can occur given the configuration of the interface itself. According to McLuhan's general medium theoretic approach, the development and maintenance of one's ethos or character in this new social sphere can be enabled and encumbered in various ways by the blog's most basic default structure. A blog is, at root, a reverse-chronologically ordered, source/author-stamped and time-stamped, text-based linguistic thread. While there seems to be something different and new about blogs because of this structure, the social and political role potentially played by blogs continues to be superficially considered. However, if commentators still hail the open-access structure of blogs without looking beyond that appearance, this is no real surprise. To the contrary, it is a trend that seems to describe well humans' relationships with technology through history.[3]

Heightened degrees of self-awareness and self-referentiality are but two consequences we can observe in a substantial portion of the human population with the shift away from oral communicative contexts, to print-oriented contexts. While McLuhan sometimes belabored the point, the basic idea is straightforward enough. In the written/textual format linguistic symbols tend to linger, in the spoken they fade. This is the primary reason why writing is a more self-conscious way of being. A slip-of-the-pen or keystroke is far less likely than a slip-of-the-tongue. Once one's words have been entered onto a blog, the fixity of the *permalink* only makes this phenomenon of heightened awareness more marked and more widespread. Awareness of the self as social actor is enhanced in the text of a blog with the understanding that one's words are then theoretically readable by the world. On blogs, private political discussions become public and create this odd, hybrid sense of private and public awareness that has recently been dubbed "publicy." The following explanation regarding publicy is from a blog created by Mark Federman devoted to McLuhan at the University of Toronto.

> Blogs are an instance of "publicy" (privacy that occurs under the intense acceleration of instantaneous communications). Our notion of privacy was created as an artifact of literacy—silent reading led to private interpretation of ideas that lead to private thoughts that led to privacy. Blogging is an "outering" of the private mind in a public way (that in turn leads to the multi-way participation that is again characteristic of multi-way instantaneous communications.) Unlike normal conversation that is essentially private but interactive, and unlike broadcast that is inherently not interactive but public, blogging is interactive, public and, of course, networked—that is to say, interconnected. There are many other aspects to, and instances of, *publicy* besides blogging, but blogging may be the most vivid example of *publicy* of mind that represents the outering of stream of consciousness or inner dialogue.

> (www.mcluhan.utoronto.ca/blogger/2003_12_01_blogarchive.
> html#10718409336)

Publicy suggests also that there is a certain degree of spectacle associated with much that occurs online today. There is, indeed, a kind of celebrity-worship, and often a kind of self-as-celebrity-worship associated with blogs. It is for this reason that news blogs (not at all immune to the reversion to publicy) represent a new form of political participation and social interaction. But, as continues to be the case even today with the cultural-lag-in-awareness regarding the socio-political significances of Gutenberg's press, we have only begun to grasp an important social fact about our emerging Digination. It is that the various modes of interaction made possible by the Internet constitute forms of life and new ways of seeing and being in the world. Participating in a news blog is a mode of existence which constitutes a new kind of political identity and consciousness. Consider the second blog tetrad drawn up in the appendix. It depicts some of the key cultural and socio-political ramifications of the "news blog."

In the previous chapter I fashioned a very simple tetrad describing some of the ways e-mail was perceived and used by my Mohawk research participants. I also described some of the unpredictable reversals or side effects of that perception and use, and the way a cycle of mutual reformation ensued between the users and the technology. Again, I append two different iterations of the news blog tetrad at the end of this book to highlight some additional issues that may be more extrinsic to the medium itself and, in the process, critique McLuhan's infamous "medium is the message" aphorism. In the chapters that follow I cut myself back to appending one tetrad per technological artifact, and weave any critique into these single tetrads in an effort to streamline the discussion. The point is that the tetrad as heuristic device carries enormous potential in helping the keen observer understand the nature and function of a technological artifact in any cultural space or moment.

McLuhan's signature statement about the relationship between a medium's content and its form continues to be read popularly as a statement that portrays content as a superficial concern. A literal reading of the statement suggests that the content of a medium is almost entirely impotent in the face of that medium's formal properties. However, it is more likely that McLuhan was playing the part of gadfly with that statement, prodding us to take more seriously the constitutive role of the technology itself in the communication process—always a cyclical, reciprocal, active and ongoing process incorporating the sender, the receiver, and the medium in between (and surrounding).

Notwithstanding McLuhan's prodding, we can get a more holistic picture of how ethos and identity might be established, maintained, and dissolved on blogs by invoking George Herbert Mead's (1934) symbolic interactionist doctrine. Mead's very holistic approach to understanding communication and cognition can help us understand some of the nuances of ethos and identity construction on blogs even if blogs represent, on the face of it, a communicative form (still) constituted substantially by the linguistic symbol. With the knowledge that Mead never lived in the age of blogs, and given that his original theory is now often buried beneath nearly 75 years of reformulation and restatement, a careful look at several of his original concepts is in order. We will then be in a position to reapply Mead's powerful ideas to the blogosphere.

Aside from the ancient Greek assertion that *virtue without an audience is unintelligible*,[4] Mead's notion of the *social role* is one of the first formal acknowledgements that the process of human identity formation may not be the individualistic, inside-out phenomenon conventional psychology suggests, and which results in a determinate (i.e. individuated) nexus of personal identity. Mead's contribution is noteworthy because he highlights the constitutive function of social life and its inherently interactive nature, where an individual is only one of many architects responsible for constructing their sense of self and identity (to others) out of the myriad social relationships they maintain. Identity formation, on this view, is a persistent and inevitable social process wherein those relationships most operative in the construction of the self typically include one's collection of significant others. During Mead's time this still meant a relatively narrow circle of family, friends, and associates, most of whom were accessed in oral, face-to-face contexts.

Mead did not explicitly address the formative role played by the mass media of his day in any sustained way (including a wide array of printed media, radio, and some film). But, there is no reason to conclude that one's experience of the world gleaned from interaction with and through these various media could not also be constitutive of an individual's identity and sense of self in a profound way. The rapidly expanding popularity of blogs (of all kinds) suggests that blogs can and will serve this function too. While not completely autonomous, the degree to which interaction on blogs forms its own kind of cultural surround seems to be unprecedented still. That is to say, aside from chat rooms (which do not tend to host the same kind of sustained political discussions as news blogs and similar news lists), people participate in blogs in a way that represents an almost total and complete mode of interaction with an unseen "public."

Mead's *significant symbol* is what makes possible the incorporation of the *social* into the experiential base that informs the conduct of the individual. On news blogs significant symbols are largely represented through the textual linguistic streams a blogger creates and observers track and comment on. Mead's attendant notion of a *conversation of gestures* was originally meant to describe the process of symbolic interchange (including everything from spoken utterances to the subtlest of unconscious bodily movements manifested by others during interaction). On blogs, the conversation of gestures is again condensed to the textual, but what follows from this central axiom of the social interactionist position remains the same: *the content of a mind is the product of social intercourse.* That is to say, *individuals* are always born out of the social environment within which they have been nurtured—incorporating whatever peculiarities that environment entails.

This is, of course, an insight of tremendous importance which leads to complexities in the social world that soon exceed our ability to parse. However, it is nothing but the taking over of the attitude of the *other* and the development of self-consciousness. This, in turn, forms the basis of Mead's notion of the *generalized other,* which comes about in infancy through imitation and solidifies further throughout the "normal" play life of young children. Now, the more seasoned student of symbolic interactionism will bear with me as I draw out some of these basic processes for anyone coming to this anew.

A fuller consciousness of the self arises when the individual (in a kind of imitation pattern) takes the attitude of another toward his/herself. In acting out his/her role of another, the individual discovers that the activities have become part of his/her own "nature." Because of this, Mead said "We must be others before we are ourselves" (1934). The French phenomenologist Jean Paul Sartre later coined a similar phrase. He said our "existence precedes [our] essence" (Sartre, 1956).[5] Out of the regular interaction maintained within this ever-widening circle of interaction there would inevitably form multiple, largely normative readings of the conduct of others. This is a recursive process that is in high gear when we are young, but which continues throughout a lifetime. In evaluating the conduct of others, the individual comes to rely heavily on that observed conduct as judgment criteria for his/her own conduct. This reflective pause we often engage in is part of the process Mead, and later his protégé Herbert Blumer, called *minding.*

Whether participating on a blog or not, social actors are constantly engaging in this process Mead described. By way of our imagination, our incessant perceptions of the world around us, and in our conscious thought, we are utilizing our attitudes (formed in large measure via others' attitudes and opinions) to bring about a different situation in the community of which we are a part. We are exerting ourselves, bringing forward our own opinion, our own meanings, and interpretations, criticizing the attitudes of others and approving or disapproving. But we can only effectively ever do that, claimed Mead, if we can call out in ourselves the response of "the community" (i.e., Mead's *generalized other*). On a news blog, with the fixed text there for all to see, these processes become much more explicit. But while more explicit and perhaps more conscious on blogs, where participation can sometimes have a sort of quasi-synchronous feel to it, one can see how the whole process might also get short-circuited without the much broader cache of non-verbal symbols (immediately significant or otherwise) that is always available to us in the face-to-face realm. Especially when invested in the topic matter, bloggers regularly respond to the running commentary as if it was bona fide social interaction. Indeed, it can be very real to participants, and this is why bloggers will tend to define themselves through their blogs. However, there are problems with self-definition in any online venue which centers upon the explosion of the "social" in our phenomenal experience today.

The relatively narrow circle of interaction that was common to most people before and during Mead's time has progressively widened as we have moved from the essentially closed social sphere of the tribe, to the rural, and then urban experiences of the farmer and fur trader, to the cacophonous sensations of the factory worker of the nineteenth and early twentieth centuries, to the increasingly media-saturated lifeworlds of the late twentieth and early twenty-first centuries. But however one chooses to dice up historical epochs, within Mead's framework, the self has always been an emergent feature of ongoing social existence in a world replete with symbols and symbol generators—no matter their form. While Mead seems to be coming from a Western perspective, he hints at wanting to proffer a trans-cultural theory of the self. To be sure, both a strength and shortcoming to Mead's thought is the sense in which symbolic interactionism can account for human experience universally.[6]

THE CONTRACTION OF SPACE/TIME AND THE DILATION OF THE NARROW CIRCLE

Ever since Negroponte (1995) coined the concept of the "daily me" much attention has been paid to network technologies and their ability to *isolate* rather than *connect* people (Discourse, 2005). One thing Mead would have been unlikely to predict is the way in which the once nearly discrete sense-making systems of his time (again, print, radio, and film primarily), would eventually, and inevitably intermix. This includes above all perhaps, the extent to which interaction online in general, and on blogs in particular, is forming an increasingly important part of people's experiential surround.

Blogs are interesting in this regard because of the way people use them. Bloggers tend to use blogs as a near-exclusive form of interaction for the persona/e they maintain there. In some sense, this brings us back to the first quarter of the twentieth century when Mead was writing about a much more stable world where relatively isolated, enclosed meaning systems were the norm. It brings us back (with a twist) because Mead was describing a "narrow circle" of friends and associates—small groups of people, almost all of whom were known first-hand by the individual. Another characteristic aspect of earlier social spheres was their finitude. With radio and film for instance, an awareness of other social realities emerged, but it was a sporadic and in no way immersive phenomenon for the vast majority of people at the time.

With the Internet and the blog we begin to see something quite different—a social sphere that is at once unseen, largely unknown, and, for all intents and purposes, infinite. We have to remember that Mead was writing on the brink of the television/electronic age. With nightly broadcasts of the war in Europe, for instance, middle class Americans very quickly found themselves immersed in a deep sea of information and were thereby compelled to incorporate a much wider range of considerations, issues, and events into their socio-cognitive repertoire. This, in turn, prompted a growing familiarity with the perspectives, opinions, and insights of broadcast personalities. It seems reasonable to suggest that in conjunction with this incorporation was a kind of alienation from the self, and with it, a growing unfamiliarity concerning one's own experience of a grounded life-world.[7]

"When things come at you very fast, naturally you lose touch with yourself. Anybody moving into a new world loses identity . . . So loss of identity is something that happens in rapid change. But everybody at the speed of light tends to become a nobody. This is what's called the masked man. The masked man has no identity. He is so deeply involved in other people that he doesn't have any personal identity" (McLuhan, in Benedetti and DeHart, 1996). Blogs seem to enhance this effect. With blogs we get the sense of real-time, or near real-time personal accounting of local and world events but without the in situ embodiment we typically associate with real-time or synchronous experience. This is, of course, not entirely new, as the telegraph and telephone (and, while then limited to one-way communication, the radio and television) pre-empted this mode of experience.

However something important here was glossed by Mead. It was due in large measure to the contingencies of his place in history. It has to do with the different degrees of sensory immersion and synesthetic integration various media often entail. This includes everything from the "lean" nature of printed text to the broad bandwidth of interactive audio-visual Internet news sites. Once again, since the world Mead described was primarily one of face-to-face interaction, that mode of symbolic exchange was Mead's paradigmatic instance (though, to reiterate once more, printed media, radio, and film certainly would have also, already been contributing something to many individuals' constructions of reality by this time). Today we know that there is not such an exclusive and direct connection between face-to-face interaction and interpersonal communication. *Mediated interpersonal communication* has become an established research area of its own (see especially Cathcart and Gumpert, 1982 for the groundwork on this). Mead's template suggests, however, that participation on news blogs might function constitutively in the identity-building process for a fast-growing part of the population today.

INTERACTIVITY, ATTENTION, AND *IN*FORMATION

We'll next consider what I take to be the three most important issues surrounding the problem of identity formation on news blogs vis-à-vis a social-interactionist framing. In no particular order of importance they are: (1) the meaning of interactivity, (2) the degree of attention (or awareness) paid to a given stimulus, and (3) the relevance of the information at hand. While these are by no means discrete issues, they should be considered individually for the sake of clarity and because they require us to consider those more ephemeral variables that constitute the social realities we create and encounter each passing moment. These issues demand a greater focus on the semantic import and the meaning and relevance of symbolic interaction (whatever form it may take) between people. How do individuals become part of a social or cultural unit? What makes one an insider? What is this process of *in*formation? These questions really do prompt a full-blown phenomenological approach to understanding social processes in digital venues like blogs. While an entire book needs to be dedicated to that kind of analysis, I can at least lay some of the groundwork here.

Nowhere in Mead's corpus does he require a correlation between one's sense of what is authentic and significant, on the one hand, and the relative information capacity (i.e., bandwidth) of a medium of communication, on the other. In other words, there is no necessary relationship between total sensory immersion (be it face-to-face reality or some kind of virtual reality) and one's feeling of what is real. This is why a series of e-mail exchanges between two people who have a lot of important and/or heart-felt things to say to one another may *be interpreted* as more real or "true" than a series of face-to-face exchanges between two people who do not. It's also why a three word text message can sometimes have more meaning than a conversation. The actual veracity and richness of the symbolic content being exchanged is beside the point. What matters is that somebody thinks it matters. This observation seems, at some level, to bring into question McLuhan's statement that the medium is the message. If the relationship between interactants is considered by those individuals to be real and true, then it makes sense that there is something in the relationship—in other words, some content—that is operative in the meaning of the message.

Erving Goffman went on to extend significant aspects of Mead's general social interactionist approach. Goffman (1967, 1973) considers human behavior by way of a dramaturgical metaphor (popularized also by the rhetorician Kenneth Burke in the 1940's). With Goffman, interpersonal communication becomes a *presentation* through which various aspects of the self are projected and portrayed. The person faced with a situation must somehow make sense of or organize the events perceived. What emerges as an organized happening/event/episode for the individual becomes that person's reality of the moment. In Goffman, what is real for the person emerges from that person's definition of the situation.

The most relevant area in which Goffman extended Mead was in considering the perceived attitudes one has of people and characters (living and dead, factual and fictive) as they operate in a person's assessment of the situation as well as their self-assessment. Here again however, as with Mead, Goffman neglected to consider seriously the potential mediated communication and, more precisely, tele-mediated information has to construct a *real* situation for someone (c.f. Kim and Biocca, 1997, for an insightful discussion of *telepresence*). Neither Goffman nor Mead gave much consideration to the potential social and psychological effects associated with differential sensory immersion and what this means in terms of interactivity, attention and the establishment of a relationship, a sense of self, and even a knowledge base.

Once again, by stressing the importance of *sensory immersion*, we are not required to make any claims about the purity, isomorphy, or naturalness of face-to-face communication. Determining those characteristics of the face-to-face mode (or any communicative mode for that matter) really is the task of an aesthetic analysis. The present work has been more of a functional analysis concerned with the phenomenological and epistemological significance of blogs and blogging. And so while there are certainly ideal contexts for face-to-face communication, such interaction need not be considered the ideal or paradigmatic case for communication universally. To idealize the face-to-face at this stage in our collective history would be to ignore a great deal of positive, practical social activity happening elsewhere.[8]

Interactivity, awareness, and the perceived relevance of information can help us understand the actual dynamics of identity construction across communicative realms. By *actual* I mean something like the pre-theoretical (i.e., objective, empirical) foundations which lead to this experience of identity. We can consider the face-to-face ideal again briefly to demonstrate how it is that one's sense of self (and one's sense of social and physical place) is established and maintained by way of these three considerations. The important difference between mediated and unmediated encounters is that, in the mediated situation, there is a reduction in physical and non-verbal cues, and, sometimes, an absence of direct feedback. It is the latter which usually raises critiques in regard to the "communicative potentials" of mediated forms of interchange.

But why should the presence of communicative feedback necessitate the reception and integration of information? Counterintuitively perhaps, we have to accept that human communicative feedback is only ever a *potential* variable subject to the design features of a medium and, as such, might often be overrated, and/or under-analyzed in some cases. If the preceding sounds like an odd question, we should then ask why most of us feel that face-to-face communication fosters the highest levels of awareness, attention, and truth when it comes to getting the correct understanding of symbolic content. There is no compelling evidence which suggests that "face-to-face" equates in any consistent way to "interpersonal" at this point in history. This may appear to be turning the whole debate on its head but, to reiterate, even though face-to-face interaction carries with it the highest degrees of sensory immersion or *sensory integration potential* (Avery and McCain, 1982), that full potential is not always exploited. Indeed, we have all experienced, at one time or another, the absent, impersonal face-to-face exchange.

According to Avery and McCain, the technologies of mass-mediation can inherently limit the *sensory integration potential* for receivers. Despite the acknowledgment that the human sensory apparatus is a homeokinetic one (i.e., self-stimulating and self-perpetuating), Avery and McCain offer a decidedly media-determinist perspective which downplays this homeokinesis, rendering it of secondary importance to the constraints of the medium. Nonetheless, they seem to accurately characterize the human perceptual system in the following passage: "Human sensory modalities are not dormant sensors waiting to be stimulated and affected by objects in the environment. Instead, the perceptual systems of the human body are searching systems, actively and constantly scanning their environment for information appropriate to the needs of the information processor" (Avery and McCain, 1982; p.30).

While the machine metaphor the authors employ has some limitations, we can extend that trope presently and recognize that the human sensorium does not simply reach out to scan the external environment. Our perceptual apparatus also attempts incessantly to tap reflexively back into memories stored through a variety of means: audio-verbal, audio-nonverbal, visual-verbal, visual-pictorial, olfactory, tactile and taste, etc. And this all points back to Mead's theory as well. Consider the case when, say, the smell of pine pitch on the way to a job interview recalls a camping trip whereby the advice of a wise old uncle resurfaces (thereby arming you with that perfect aphorism to win over your interviewers and land that brilliant job). These bits and pieces of information and opinion can be triggered by the synergy of a whole range of sensory modalities in strange places—and at the most unpredictable moments. This is why we need to start taking mediated interpersonal communication so much more seriously, with e-mail, texting, instant messaging, and blogging representing four of the most popular forms of "intimacy at a distance" (see Horton and Wohl, 1956 for an initial discussion of intimacy at a distance and parasocial interaction). Thus, in addition to the total sensory immersion described by Avery and McCain, we must consider our knowledge of others and the content of others' knowledge as key variables in the sense-making process. In the same way that favorite uncle's great advice can resurface at the right moment, the words and deeds of distant friends, unseen experts, and even archrivals often provide us with a rich supply of symbolic resources in coping with the exigencies of daily life. Recognizing how such mundane instances of interaction provide the building blocks for identity and action, we begin to see how relationships constructed out of consistent and focused reading on (and perhaps periodic commentary posted to) a blog discussing Massachusetts Senator Scott Brown's rumored aspirations to the U.S. presidency might prompt more of a sense of reality, more genuine feelings, and provide a more compelling sense of self than any face-to-face relationship can.

This is not difficult to imagine. Given the topical interests, the hopes, the fears, even the mood of a particular individual, one of the most rudimentary, one-way, asynchronous, low-bandwidth communication channels (i.e., hardcopy text) may prompt the highest sensory integration potential. The reason for this is because, even though mediated encounters tend to alter the sense modalities in various ways, communicative participants can either "supply the missing sensory data from their past experience or 'know' the object in question based on 'incomplete' sensory data" (Avery and McCain, 1982; p. 30). The truth is that many face-to-face interactions can and do often fall short of the sensory integration potentially available in them. But we should not conclude from this that some less-than-potential which exists then translates into a communicative shortcoming of whatever medium is interposed. Once again, the relationship between participants, the social situation, and the topic of discussion also determines and constrains the extent of sensory data available, not merely the medium of exchange. A given interaction may not exploit the full symbolic potential that exists in a given channel of communication.

Conversely, where sensory potential is formally lacking, it may be created. There is an emergent, spontaneous aspect to meaning-making in human communication. This is often reported between family members, or best friends and lovers who describe feeling emotionally "touched" or intellectually captivated by even a misspelled, poorly punctuated, grammatically awry text message. In that case, it is not the medium doing much of the communicative work, but rather the message—it is the nature of the relationship that is operative in both the epistemological (knowledge or sense-making) and ontological (reality) functions. In sum, the potential for full sensory integration, and the attendant availability of feedback in a communication system does not guarantee that meaningful interaction will be experienced by the participants themselves, nor perceived by them with regard to others (as when two people engaged in conversation in an "intimate" dinner setting manage to talk past one another for the entire evening).[9] Of course a similar kind of "tuning out" regularly occurs during telephone calls, via e-mail, and on blogs when one of the interactants is glossing a paragraph, skipping to the sign-off, or merely waiting for the other person to finish. But again, that *tuning out* is not necessarily a function of the medium itself. One's character (positive or not) can be constructed and maintained with similar robustness in and through a variety of media. As a practice and as a communicative form, blogging is situated between a variety of different tensions including those generated by the features we typically associate with orality on the one hand and textuality on the other.

It has been demonstrated in the second half of this chapter why a monolithic emphasis on the structural features of news blogs might lead an observer astray in attempting to ferret out the mechanisms of identity construction taking place there. The preceding hopefully reveals why even fairly rudimentary tele-mediated forms of communication, like that which takes place on news blogs, can seem so compelling and so real. In the same way our homeokinetic perceptual systems will do their best to fill in the blanks during phone conversations with, or letters from, significant others, they often have no *reason*, no *urge*, no *felt-need* to do so during face-to-face interactions with others less significant to us. This just seems to be an extension of a normal and often understandable human bias toward familiarity (for better and for worse, "birds of a feather" have always tended to be attracted to each other).

Of course, this homophilic tendency can have quite deleterious effects when put into play in a social space like a news blog, which is presumably dedicated to the dissemination of information and the interplay of opinions and assessments of that information. Recalling Norris's (2004) study of online communities, we learn that interaction tends toward the homophilic in places like blogs because of the ease with which people can enter and exit these digital domains. The avoidance of cognitive dissonance is also a common default behavioral tendency in humans and, indeed, it is often easier to simply "exit than to try to work through any messy bargaining and conflictual disagreements within the [blog]" (ibid, p. 33).

Like Norris (2004), we should be concerned with the loss of face-to-face interaction. But it should not be a concern stemming from our necessarily becoming less natural, or less real while online. This was clearly a concern for many of my Mohawk friends using e-mail, but it should not be a categorical concern. As the preceding analysis has hopefully shown, we should be more cognizant when venturing into the online worlds of words and images because so many of us do so for so many different reasons—too many, in fact. Like a hammer, then, a website, an e-mail message, and a blog should all be thought of as certain kinds of tools. And so we need to stop thinking about communication media as neutral vehicles that we use to transfer thoughts, ideas, and messages. Especially in this age of the image and the tele-mediated word, we need to start thinking about communication media and the messages embodied by them as occasions for people and institutions to create selves, mold identities and construct entire realities—this is the nature of Dignation.

With McLuhan's tetradic analysis we see how the news blog may systematically enable and disable certain features of interaction. With Mead, we have seen how the process of identity construction that takes place on blogs can maintain such a compelling sense of reality even though the symbolic bandwidth (which is theoretically at its height during face-to-face exchange) is compressed down to the linguistic/textual code there.

> Many believe that any erosion in the traditional face-to-face sociability and personal communications of gemeinschaft in modern societies represents a threat to the quality of civic life, collaborative social exchanges, and the community spirit. Whether the Internet has the capacity to supplement, restore, or even replace these social contacts remains to be seen. As an evolving medium that is still diffusing through the population it is still too early for us to reliably predict the full consequences of this technology (Norris, 2004; p. 40).

Blogs continue to stake out significant pieces of real estate on the Internet. If these arenas don't get us any closer to some idealized notion of face-to-face interaction, they do seem to allow us to get, in effect, *closer* to others who inhabit the same cultural unit, occupy the same interpretive/speech community, or have similar affiliations and affections as ourselves (can we, in fact, talk about the myriad others we "meet" on news blogs in terms of Mead's *significant others*?). The political and civic functions of news blogs may not be what they appear.

Due in part to the formal structure of the interface, and in part to the ease by which a visitor can arrive and exit, blogs may not enhance the interplay of diverse political perspectives in the way they have been billed. However, news blogs will likely remain an important part of many people's everyday lives because they allow us—and we seem to use them—to reach others like ourselves. For this reason blogs will continue to foster a profound sense of community within select political and social groupings. Next, we'll roam around the Internet a bit more broadly to consider the nature of the web of information individuals confront in their (mostly) solitary investigations. It is this primarily individual relationship between human and machine sustained today that does much to help constitute our Digination.

NOTES

1. "News" is a slippery term these days (due in no small measure to the expansion of its semantic import via blogs). However, for our present purposes "news" will refer to any journalistic information that could be used by the public to enlarge and enhance awareness of the world and, by extension, enable more informed civic and political participation. This definition does not, admittedly, do much to bracket the twenty-first century phenomenon of news; however, it is primarily a descriptive and practical definition as the present study is largely exploratory and not intended to be evaluative at this stage.

2. Technological determinism implies that the entire process is now largely out of rational control—coupled with market logic and one has a self-prodding system (c.f. Altheide and Snow, 1979). Web designers are compelled to keep pace with the technical capacities of the technology. This is why we will continue to see the Internet pushing the limits. But economics also underlies part of this story. Indeed, we are moving toward a total sensory-immersive environment centering upon image and sound that is no doubt due in part to an implicit business agreement between Microsoft and Intel. When Intel brings to market a new 3-Gig processor, Microsoft reorients to the production of new, updated, and "improved" software that utilizes all of that processing power and new bandwidth capacity it entails. As Microsoft brings to market the latest power-consuming operating systems, office suites, Internet browsers and media players, Intel is compelled to develop a chip that allows for the next innovation in software. This in turn allows Microsoft room to add more color,

sound, and interactive capacity to their products, which prompts Intel to perfect the newer 3.5-Gig device, and so forth.

3. The idea that a technology (like writing) can exert causal force in a manner that prompts actual changes in the user such that people "give in" to the various structural demands, logics, or representational dictates of the medium was first articulated in a formal way nearly 2500 years ago by Plato in his dialogue *The Phaedrus*.

4. A paraphrasing by Nietzsche, actually, referring to one of the ancient Greek philosophers. It is probably Aristotle, as this same notion is evident throughout much of his work—most notably discussions of ethos and pathos in the *Rhetoric and Poetics*. In partial criticism to Plato's idealism (the groundwork upon which most individualist psychological theory is based), Aristotle makes a point of demonstrating that actions are always in an important sense public (i.e., intended for others to see or, at least, with others in mind). The possibility of non-publicly-displayed social behavior follows from this.

5. Some nineteenth and twentieth century philosophy also addressed this tension between the individual and the collective. Hegel's *master/slave dialectic* and Sartre's *"look"* are two notions suggesting that there is some inherent creative potential in what might otherwise be described as a decidedly *re-active*, multi-person system. But neither of these thinkers drew out what I think is the most important aspect of such a conclusion: the idea that we (collectively) make our worlds. Hegel simply posed an inevitable, trans-cultural progression toward some total, unified self-awareness (essentially, God becoming self-aware through man again). Sartre, on the other hand, was compelled to focus his inquiry upon the manifest lack of foundation threatening all attempts at valuation and, hence, the apparent meaninglessness of human existence. The human challenge for Sartre was to make ourselves. The human being is compelled to imbue their world with meaning and thereby avoid the chasm of oblivion. I think Sartre never saw much sense in drawing out just how it is we do that. He felt no need to make prescriptions since that very act would function to impose yet another foundationless construct. But that is precisely what we want to know: *How* we make meaning for ourselves and the objects around us, *how* we make a place in the world. Mead helps somewhat with this enterprise. If we can accept the possibility that the social world is an intricate fiction, then the difficult work is done. This allows us next to scrutinize the sense-making systems themselves without the heady criticism. When we stand back and look at these constructs, we discover that the world we inhabit is a shared creation, an elaborate, mutually sustained web of interaction which we must constantly assess and reassess to varying degrees of depth as we progress as individuals and as a species.

6. There is a sense, in Mead, that *thinking about the self* is everything but that since when someone says they are thinking of themselves, they are more accurately thinking about what others think of them, how others see them. This seems to have also been Aristotle's position on the matter.

7. As one local journalist has put it while referring to the phenomenon of the cineplex theatre, the loss of familiarity with the grounded life-world is in part a result of certain tele-media basically playing a surrogate and, in some cases, impoverished function. These media perform well in as much as they "convey a broad locality that

refers to deep, uncontrollable urges: the arena of dreams, of public life, of desire. We gather, we watch, we assimilate (or at least try)—a timeless ritual. God is dead, we don't know our neighbors, and the connective tissue that binds individuals in our society is either absent or needing repair or complete reinvention" (Keepnews, 1997). Now, whether the effects are as bad as all that is questionable. What seems incontrovertible is the suggestion that some basic underpinning has been torn away.

8. Such idealized claims regarding face-to-face communication continue to be made (see especially Bavelas, Hutchinson, Kenwood and Matheson, 1997).

9. This is roughly the distinction made by Kenneth Burke between *action* and *motion*. The latter (like the dinner conversation noted above) characterizes interaction that merely satisfies the mostly tacit structural features of discourse (i.e., my answering your queries with the appropriate nods and/or verbal acknowledgements or your responding to my statements with the appropriate "hmms," "aahs," and "uh huhs"). This mere filling of slots can characterize face-to-face communication as much as any other form. When this is the case—when participants are only satisfying the formal syntactic criteria of discourse, there is good reason to believe that the majority of the sensory modalities are, in effect, turned off, or at least "tuned out." Hence, sensory immersion becomes a function of attention and identification as much as medium.

Chapter 5

Information, Interactivity, and the Denizen of Digination

I asked my Communication Research class (mostly seniors, a few juniors) this afternoon what the "big media news" was today. No one responded, so I told them it involved some media giants . . . and gave them a couple more hints.

Not one of the 24 knew of, or had even heard about, Comcast and GE/NBC (the proposed merger). A typical comment was, "I don't really read the news that much." I asked how many were PR concentration (about 40%) and how many were broadcasting (about 35%). I suggested that perhaps they might want to follow the news a little more regularly.

This reinforced my personal, anecdotal, unscientific belief that newspapers are not losing news readers to the Internet, but rather we're raising—and graduating—students who are almost totally unengaged with the world beyond the latest sports and entertainment headlines. (D3)

This commentary offered by a mass media professor at a liberal arts college in the Boston area highlights the irony of the modern media age. Her assessment might even be a bit too critical of her students however. Indeed, their lackluster performance during class discussion may be more the side effect of an underlying and largely invisible environmental change long underway. It is the connection between a culture-wide shift from words to images, and the subsequent loss of meaning and meaningful content that hangs in the balance.

The United States has one of the most advanced, evenly distributed, open and accessible communication systems on the planet. While similar digital infrastructures have been built with similar transformations occurring in other regions, American information technology is setting much of the pace—and the template—for the rest of the industrial and post-industrial world with regard to the way digital communication technologies and systems are being deployed and utilized.[1] The problem is that Americans continue to be one of the most poorly informed populations (in terms of diversity of opinions/ sources, depth and breadth of knowledge, etc.). Our paltry math, science and civics scores surely account for some of the lag, but something else is at work here. And, perhaps ironically, the proliferation of personalized information services, photo news galleries, computer simulations, and a host of interactive media links on commercial internet news sites have been hailed recently as one remedy for these troubling statistics.

But this American-centric trend may not continue for long. In 2005, after massive communication infrastructure investments, the nations comprising Western Europe came to represent the largest concentration of *netizens* in the world with more than 300,000,000 people connected to the Internet, many seeking the same conveniences enjoyed by their American counterparts. This chapter examines the relationship between technological capacity and usage patterns on several of the leading internet news sites from a medium-theoretic perspective. I'll argue that as the Internet becomes more technologically sophisticated, a proportionate, though inverse trend in the *epistemological sophistication* of its user base will be inevitable. I'll close the chapter considering a few of the implications this rule and this trend holds for the future of a global citizenry.

While I couldn't resist opening with that commentary from a colleague, the motivation for this chapter actually springs from my own classroom experiences over the past fifteen years at a large state university in New York, a small private liberal arts college in Northwestern Pennsylvania, a school specializing in communication in the heart of Boston, and a small liberal arts institution just outside the city limits. I was immediately surprised by the lack of what I considered to be key civic, political and geographical knowledge many students coming into my classes exhibited. As might be expected, local and national politics, current events, and international news were the leanest content areas. What I took to be basic knowledge necessary for productive work in my upper-level media theory and political communication courses for instance, a majority of students found irrelevant and obscure.[2]

No direct causal claims associating individuals' lack of knowledge of and about certain issues, on the one hand, and their use of the Internet on the other, are being made in what follows. Clearly, the American educational apparatus is not yet what it needs to be. I want to propose, however, that certain structural features of the Internet functioning alongside common Internet usage patterns today may play a significant role in exacerbating ignorance and/or misapprehension about both domestic and world affairs, and even non-participation in social, environmental, and political issues.

I investigate one significant aspect of a very complex phenomenon: the progressive move away from the printed word to the image as the primary and preferred means of human symbolization. It is important to note that this last clause is purely descriptive. Counter to the prescriptions of many business and world leaders today, I would caution against a wholesale buy-in to the notion that giving every child on the planet a laptop with an Internet connection will solve their problems, let alone ours. Rather, as I hope the following discussion makes clear, the shift away from the word to the image is a predictable stage in the evolution of digital communication technology, and so a natural part of the process of Digination. To be clear, the move from linguistic information to image-based or pictographic information reflects biological patterns of dissemination and is a sort of natural byproduct of modern digital media design and consumption. Apple's *iPad* is emblematic of this shift, with a platform that is designed to be far better at allowing the user to consume content, and image-based content above all. Odds have it that the next generation *Kindle* and *Nook* will also follow suit to stay viable in the marketplace.

I'll discuss first some of the empirical work informing my observations. This is followed by a medium-theoretic inquiry into the significance of the data. I then reflect upon a few potential consequences that might be associated with a popular move toward the Internet as a primary news source and a concomitant shift toward the image (both still and moving) as the operative informational unit.

While my students sometimes have trouble getting their heads around it, the supremacy of the image is a very recent development in media history. The culture of the word began its rise with the invention of the phonetic alphabet about 5000 years ago, and an awareness of the power of the word was first articulated in a formal way nearly 2500 years ago by Plato in his dialogue *The Phaedrus*. In that document, Plato—extending some of Socrates' own opinions on the subject—suggests that the steady proliferation of the written word would be accompanied by a progressive bias toward the eyes to acquire information.

Jumping ahead to the mid-nineteenth century with the advent of the *Daguerreotype* (the first photographic technology), we find evidence for another vision-oriented media shift in the world, and in the human sensorium, and subsequently, in the way individuals make sense of their experiential worlds. The social impact of the image at this time was generally known. Photography was transforming the way people see and interpret the world around them. However, pundits and publics alike maintained high hopes. Indeed, for an extended period between the Civil War and WWI, the photograph was even considered "legal reality." Early photodocumentarians believed they could represent the world *as it was*, or "in fact" to the masses. The argument was that people would begin to see the true conditions of existence (in war, famine, poverty, etc.) and be moved to action by the images they saw. The expression "seeing is believing" was becoming a kind of truism.

The twentieth century records a series of smaller revolutions in imaging technology: from silver nitrate film emulsion to motion picture film and cinematography; from fixed camera placement to movable track and steady-cam systems; from linear analog videotape editing to computer graphic imaging (CGI) and non-linear digital editing. Today we know that practitioners have nearly complete control over the way they want *reality* to appear.

Similar to Plato's critique of writing which was alleged to take away our capacity to remember by prompting an over-active, and often inaccurate imagination, we can conceive of modern imaging technology as having attenuated our *ability to make sense* and perhaps also our very *need to imagine*. This seems to hold if we consider the manner by which these two modes of symbolization—the word and the image—are processed by the human cognitive apparatus.

Today neurologists, cognitive scientists, and linguists are mostly in unison when they say that humans process the written and spoken word (linguistic information) in the neocortex and several regions of the left hemisphere. We seem to use both innate and learned logic and rationality in the language interpretation process. As Neil Postman (1985) generally remarked, we have proven ways of making sense of linguistic statements. We learn a given vocabulary, along with the rules of grammar and punctuation so that when we see or hear a sequence of words we can get to the basic sense, or nonsense, of what's been said.

But the environment of the word has now largely moved to the periphery. If the apex of *literate culture* in the United States was shortly after the turn of the nineteenth century, it has been usurped at the turn of the twentieth by the photo image, the filmic sequence, and the pictographic representation (c.f. Meyrowitz, 1985; Postman, 1985). Today there is also common agreement among cognitive researchers concerning the way we process image-based information. The consensus is that this processing is less rational, less logic-based, and seems to generate activity in the paleopallium, or limbic system, and regions concentrated in the right hemisphere; areas generally regarded as the creative and emotional seats. It is worth noting, without being aware of the latest research on the topic, that Marshall McLuhan opined early on, that "we have not the art to argue with pictures" (1964, p. 231). With the science behind us we now know that McLuhan may have been making much more of a naturalistic claim than he might have imagined.

The latest Nielsen data suggests that American children, ages 2–11, watch about 3 hours of television every day. And for the past decade the American Pediatric Association has strongly recommended that parents prohibit children under the age of 2 years from watching any television whatsoever. The argument is that at this crucial stage in brain development the linguistic centers need to be activated, and that frequent, long-term image processing does not allow for adequate growth in those key areas.

I think McLuhan was engaging in some of his famous hyperbole when he said that the content of a medium is like a juicy piece of steak that distracts the watchdog of the mind. However, this and other *McLuhanisms* were probably more akin to prophetic projections or rhetorical devices he employed in order to get us thinking about the powerful influence exerted by the primary media of an era. Besides, we should be able to forgive McLuhan for being enamored by his own construct given that he became something of a spectacle in his day. When he suggested that *the medium is the message* he was making a general point: that substantial shifts in the primary medium—like the moves from orality to writing, to mass-produced print, to electronic and digital systems—were the events, if sometimes protracted, that most succinctly defined historical epochs and the people who live in them. McLuhan was offering a statement about sweeping changes in the way humans make sense of their environments. I'll say he was largely on target, and then add another argument to the mix. As I suggest in the chapter on blogs, we have to sometimes look beyond the formal features of a medium. We need to consider the content as something that can also be *operative*.

Earlier, I pointed out a generally accepted historical fact. At the turn of the nineteenth century in the United States we had reached the highest literacy rate per capita (Shapiro, 1992; International Trade Statistics, 2002). That era also boasted the highest level of political participation.[3] However, these facts likely stem from more than there simply having been an unprecedented number of newspapers and newspaper readers in circulation. Indeed, what's often missed in this story is the attendant fact that the press was, for the most part, a partisan press. Journalists were, by and large, overtly biased, opinionated, and explicitly ideological in their reporting and editorial work. To be sure, the partisan press in the United States at the end of the nineteenth century was less concerned with the relaying of brute facts, and more intent upon telling good stories via opinions and arguments surrounding the facts. The next question is: why did the partisan trend fade? This question prompts us to reconsider the notion of a weak or soft kind of technological determinism in relation to a kind of hyper-utilization in the population.[4] Hyper-utilization, in turn, prompts us to consider content.

In the 1880's, when the first powered presses went into operation there was an almost immediate shift away from the production of papers as the central problem. Indeed, as with most forms of mass production, a consumption bottleneck was the direct result. As the first presses went online the production of readers-qua-consumers became the most pressing concern for the owners and operators of newspapers. To reach the broadest numbers and antagonize the least was the new unwritten rule in the newsroom and ad office. News needed to be aesthetic, and cheap. The *new press* and *yellow journalism* were two expressions that entered into industry parlance to refer to the way in which this new method of broadcasting information (the mass produced daily) created a mass audience which, in turn, prompted a new form of writing. Today we have the following popular notions of what *news* entails. According to *American Heritage* the definitions of *news* are as follows: (1) Information about recent events or happenings, especially as reported by newspapers, periodicals, radio, or television; (2) A presentation of such information, as in a newspaper or on a newscast; (3) New information of any kind: *The requirement was news to him; (4) Newsworthy* material: "a public figure on a scale unimaginable in America; whatever he did was *news.*"

The fact-based, time-sensitive, more objective journalistic style that was born with the telegraph as a necessary way of condensing transmissions to manageable levels was later officially codified in newspaper writing. Scientific objectivity became the new standard, with a dogged focus on the immediacy of the present. The new writing style de-emphasized political discussion and historical context, accenting instead the new and the now. Focusing American Heritage's definitions, this new style of news can be described as a punctuated sequencing of discrete facts, with less opinionated writing, and a more "watered down" or more broadly acceptable kind of content.

Part of this new aesthetic function was the "consummatory value" publishers demanded from their writers. Short news items needed to be understandable in and of themselves; easy to comprehend after a quick scan, and enjoyable (Schudson, 2003). If the *Daily Graphic* (1873) was the pioneer in this regard, Gannet's *USA Today* was the first modern newspaper to take this logic to the next level in 1980. As will be discussed shortly, however, a consequence of this change may have been the creation of a less active audience—both in terms of the work they need to do in decoding content, and subsequent actions they may engage in after processing it.

Jumping ahead once more to the mid-1980s we have record of another important media shift that can be described as a kind of hyper-utilization. Ted Turner's Cable News Network begins broadcasting on June 1, 1981. In similar fashion to the search for readers that was spawned by the powered press, the search for and even the creation of news becomes CNN's primary concern. With the 24-hour-news capability comes the need to fulfill the promise: *"all news, all the time."* But the public also heeded that promise and soon came to expect just that—new news with every new tuning in.[5]

Yet CNN's felt need to get the scoop is not their problem alone. The constant expectation, then demand for new information that developed over the next several years helps spawn the *spin doctor* during the 1984 presidential election campaign. But this new media personality did not do journalism, or news per se, at least not as traditionally conceived. As opposed to the explicit partisan nature of news at the turn of the century, these new purveyors of partisanship wrote, spoke, and discussed in veiled, indirect, and uncertain tropes. Their language was obscured by the manner in which it was designed—via innuendo, esoteric allusion, hyperbole, oblique reference, and strategic ambiguity.[6] But by the mid-1980's CNN knew it had to keep talking. Because of the network's unique devotion to news, Ted Turner's nationally recognized spokespeople had to fill the new and vast "news hole": that broad space that now existed between one sponsor's paid time and the next.

We need to understand that CNN was a unique institution when compared to the other major television networks at the time. And today NBC, ABC, CBS, and FOX still have a different issue to contend with. Given time constraints imposed through commercial dictate and strict revenue targets, network content providers are seldom able to link even the most important stories to the relevant past or to the ebb and flow of social, political, or natural history. As the value shifts were occurring in the news organizations Mirtoff and Bennis (1989) critiqued the news broadly along these same lines, suggesting there is "no connecting thread, overall context, or historical perspective provided that would help the viewer, reader, audience, etc., make any sense of the larger pattern of ideas, images, etc., assuming that there was one" (p. 50). MIT linguist and social activist Noam Chomsky has pointed repeatedly to the "requirement of concision" as a fundamental problem that greatly reduces the relevance and quality of news today. With all of this we have thus returned, in a very real sense, to a telegraphic-style of news reporting at the beginning of the twenty-first century for the same reason it existed at the beginning of the twentieth. There is just no time.

It seems that this trend toward a kind of *concise incoherence* based in changes in the technical features of the news media over the last century must shift into overdrive as electronic and digital optic transmission becomes the norm. These patterns of representation continued to be adopted by virtually all of the mainstream news organizations over the last three decades to become what it is a today—a firmly rooted cultural expectation. I argued in the previous chapter how blogs are moving quickly from word to image-centric formats, and now the

140 character Twitter feed is another kind of typographic throwback that I think captures the essence of this efficiency ethos. Indeed, Twitter and its ilk may, ironically, represent a further debilitation in the user base with respect to their ability to parse a legitimate (i.e., long form) typographic mode in a push for communicative efficiency that began with Morse's telegraph and has led us to a broad-based hyper-utilization of image-centric news and information forms.[7]

So if CNN provided us with obscured interpretation of the facts, the other television network's references to "The New War," "The Showdown in the Gulf," and "The War on Terror" become what one media critic has described as *news McNuggets* (c.f., Deenan, 1991), short, narrow stories embedded between colorful imagery and graphics.[8] There was a glimmer on the horizon however. The move to the Internet promised a different kind of news. When CNN invested heavily in an online presence in the mid-1990's they did so thinking that their already vast "news hole" would become virtually bottomless. Today this holds true—sort of. My research revealed that, in actual practice, so many Internet news consumers barely scratch the surface of the content the mainstream (and alternative) sites they frequent have to offer.

Supporting McLuhan's claim that, "[i]n the name of 'progress,' our official culture is striving to force the new media to do the work of the old" (1964; p. 81), Jay Bolter and Richard Grusin, in their book *Remediation: Understanding New Media* (2000), define the concept of remediation as "the representation of one medium in another." The authors suggest this to be a "defining characteristic of the new digital media." Bolter and Grusin's premise is that all new media take over and re-use existing media. News on the web incorporates formal aspects of television and newspapers. For instance, an Internet homepage presents its most important information at "first click"—that is, immediately visible "above the fold," as Bolter and Grusin put it, in the same fashion newspapers are formatted to present to lead stories in the top half of the front page. The key difference here is that commercial news sites pile on a host of "interactive" media links in their opening pages. Picture galleries, streaming videologues, and computer graphics simulations are widespread. The argument that these features are not merely window dressing around the real/serious news content finds support in my research data. Respondents explicitly name the photo/picture gallery as one of their preferred modes of news.

In the introduction to this chapter I noted that the Internet and the proliferation of personalized information services, photo news galleries, and computer simulations has been hailed recently as one remedy for the many troubling state of affairs around the world. Among those who have recognized the problem—a growing ignorance and apathy in the country— and made hyperbolic claims about the Internet, are two US presidents and a host of primary and secondary school administrators. More and more parents, teachers, and politicians are joining the push to integrate the Internet into young people's everyday lives.[9] I am suggesting that while their recognition of the problem is commendable, their remedy may be more a prescription for cultural disaster than any sort of cure. And yet, this is a core edict of our Digination—the mantra reads: *newer = better.* It would be helpful in all the tumult and therefore comforting to know that we could depend on such a clear and straightforward course of action. Unfortunately, we may very well discover that perfunctory adoption of the latest and greatest thing is not always the best way to proceed. I like McLuhan's take (published posthumously) on this issue.

> At what point is man [sic] going to recognize that this power of innovation may have to be restrained and that just as economically it may not be desirable to grow indefinitely, so that technologically it may not be necessary or desirable to innovate indefinitely? We're the first culture in the history of the world that ever regarded innovation as a friendly act (McLuhan, 1997)

Again, one of the big reasons why so many people are excited about the Internet is that it really can deliver on the promise of an almost infinite amount of information. No wonder people interested in the potential educational and research applications get so excited. Of course, this sort of misses the point. To reiterate, a central concern in this chapter is how people today go about getting news about their world. The Internet promises us, at least in theory, a bottomless "news hole": endless news and information at our fingertips. But while the promise can be backed up, I want to suggest that the problem is an ongoing confusion between theory and practice. Like so many other surveys to date on the topic, my findings show that people want to get their news fast, easily, and on their own time.

And here's the rub. There is often a very wide gulf, a kind of cognitive-behavioral disconnect, between the way so many of us like to think about our news and the way we actually go about obtaining it. The fact of the matter is that people just don't spend the requisite time getting their news and information—whether they get it from the television, which is structurally unable to deliver the appropriate depth, or the Internet, which can deliver but is often unable to do so due to the user's real or perceived time constraints. This leads us back to the issue of hyper-utilization. My academic, professional, and household samples illustrated this clearly enough. In response to a question asking respondents to list their favorite features of online news, the first three most common answers deal with the time-saving issue: *"convenience," "constantly updated,"* and *"easy access."*

This seems to support my tentative hypothesis: as the Web becomes more sophisticated (in technical terms), Internet content will become less sophisticated (in epistemological terms). I'll flesh out this emergent axiom and the basic distinction underlying it shortly. First, I need to mention that part of the issue here concerns the way so many users report being interested in getting in and out of websites quickly. It is little surprise that the fourth most common aspect of many respondents' favorite news sites at the time were the image-oriented features (like USA Today's *"Day in Pictures,"* and *"Photo Gallery,"* The New York Time's *"Interactive Feature"* or CNN's *"Graphics"* links). But let us be finally clear about the distinction I am drawing between what is structurally possible and what is bearing out in practice. That the Internet may contain all of the information necessary to build knowledge structures that can potentially solve most or all of the world's pressing problems is not enough in and of itself. This point holds for any kind of database, whether it's made of atoms or electrons: information can do nothing if it remains untapped—if it is not seen or heard, critiqued and compared and synthesized as knowledge.

And here we come to the crucial distinction between *Knowledge* and *Information*. The first three entries in *The Oxford Dictionary of the English Language,* 2nd *ed.* for the term *Knowledge* are: (1) expertise, and skills acquired by a person through experience or education; the theoretical or practical understanding of a subject, (2) what is known in a particular field or in total; facts and information, (3) awareness or familiarity gained by experience of a fact or situation. And now consider the first three *OED* entries for *Information*: (1) Knowledge derived from study, experience, or instruction, (2) Knowledge of specific events or situations that has been gathered or received by communication; intelligence or news, (3) A collection of facts or data: *statistical information.*

It is interesting to note that aside from the third entry under *Information* the two words are considered almost perfectly synonymous. It seems that proponents of the Internet listed earlier—the everyday and the influential—have accepted this conflation that really has become a somewhat outmoded linguistic convention in the digital age. Granted, the publication of the third edition of the OED (last promised before the end of 2010) might update these entries to amend some of this confusion. However, the point remains that if the editors of news agencies choose to include the punctuated tidbits that are common in many of the nation's most popular dailies and TV news specials on their Web spaces, then news and information remains just that—unprocessed (even unprocessable) raw material. This may be especially the case where picture galleries, short video segments, and computer simulations are the preferred content (both from the producer's and the consumer's vantage).

I've already detailed some of the arguments that suggest why pictographic information is less amenable to formal analysis and sense-making processes. But still, another phenomenon may be at work.

Gumpert and Cathcart's (1985) notion of the *media generation* should also be taken seriously here. Recall their four contentions: (1) there is a set of codes and conventions integral to each medium; (2) such codes and conventions constitute part of our media consciousness; (3) the information processing made possible through these various grammars influence our perceptions and values; and (4) the order of acquisition of media literacy produces a particular world perspective which relates and separates persons accordingly.

Extending Gumpert and Cathcart with respect to the formative properties of media, I'll suggest that the world perspectives fostered by immersion in a primarily visual symbolic environment results in a less coherent, less critical, and ultimately *less knowledgeable* population.[10] Gumpert and Cathcart's fourth point is particularly crucial to the idea of a "media generation" or "cohort." One thing that was abundantly clear in my questionnaire data is that the college students (those most apt to fall within the television/image and computer/iconic literate generations—i.e., born in the mid to late 1980's) were somewhat more apt to visit and bookmark sites that put a lot of energy into their "interactive" pictographic content.[11] So extending Gumpert and Cathcart's early take on this issue, it seems that the attractiveness and familiarity of the image links is what allows these younger users to stay within the limits of their shortened attention spans or "twitch times." The doing of news today is certainly a pursuit beholden to and therefore delimited by a kind of *attention economy* (Simon, 1971).

My second hypothesis, then, is really just a logical extension of the first: that there will tend to be an inversely proportional trend in the relationship between the technical sophistication of the Internet's user base, and the epistemological sophistication of the knowledge those users obtain and integrate into their respective lifeworlds. This is, of course, in direct opposition to the message being pedaled in recent advertisements by major service providers (with my favorite catch-phrase being promoted not long ago by AT&T: "The Internet makes you smarter." Perhaps this is one point where McLuhan seriously misinterpreted the possibilities of computer technology. My guess is that he probably did not take seriously enough the impact of certain kinds of content when he said things like "[i]n an electronic information environment, minority groups can no longer be contained or ignored. Too many people know too much about each other. Our new environment compels commitment and participation. We have become irrevocably involved with, and responsible for, each other" (1964, p. 24). If perfunctory involvement in other's business is part of the new ethos of social networking, the balance of insight simply has not borne out. The decontextualized nature of so much of the news users grab from the Internet today has not yet resulted in any appreciable move toward a more informed, more knowledgeable, more aware, more participatory, or more responsible group (Griffin, 2007).

Thus, here again we find an almost monomaniacal form of hyper-utilization at work. Throughout most of the 1990s Internet users were very often given the option of choosing the "*text-only*" or "*frames*" version of a website. This is rarely the case today. The result is that anyone possessing hardware and software much more than five years old is precluded from entry onto the information-superhighway. The text-only option was a hold-over designed to accommodate the text-based DOS browsers (like the original *Gopher* programs on BitNet) that did not support any graphical content. In keeping with the argument set forth earlier in this book, an argument supported generally by Eisenstein (1979), McLuhan (1962), Postman (1985, 1992), these less technically sophisticated text-based systems were the only real guarantors of more sophisticated content (in terms of depth and breadth of information, diversity of sources, etc.). As discussed in the previous chapter, the "conventional blog," if I might be so brazen to try and deploy such a term in this moment of mind-numbing change, is the kind of media form that might be able to deliver such depth and breadth when read bottom-up. However, the fast-shrinking collection of text-based Internet sites that are not built on the blog's reverse-chronologically ordered format do not require the latest processors, software, or connection speeds. Most can still be accessed and navigated efficiently with a late-80s era 386 machine running through a 14.4K BPS modem in DOS mode. Literally thousands of text-only news/information-oriented web sites still exist, but such an approach is now considered substandard from both technical and marketing perspectives. No "serious" news server can hope to respectably operate in this manner today.

Therefore, as more of us go online, more of us will unscrupulously put ourselves into a position of trading knowledge for information (where that distinction contains a difference that makes a difference). Is this also a part of Digination? The tendency to progressively move toward a kind of passive knowledge acquisition that is more akin to a brute, and somewhat mindless kind of data gathering? *Infomania, infophilia,* and *infotropism* are the contemporary ailments many of my closest friends and colleagues joke about. But it may be no joke. As Plato and certainly his mentor Socrates might observe were they around to see it all, massive caches of physical and digital files should not be confused with knowledge or learning—or wisdom for that matter.

And so regarding McLuhan's hopes about the potential for increased and irresistible participation in the global village, what many perceive as interactivity is more likely just a new form of thinly-veiled passivity. This is most obvious in the widespread use of personalized news services that send along items based upon a key word search. The mechanics of these processes will be thoroughly unpacked in the next chapter. For now, we can say this: The Internet is predicated upon the idea that events in the world are discrete and clearly delineated/earmarked as certain kinds of news with clear descriptions and associations. These systems can structurally deny users participation in the association process. The underlying logics of proprietary algorithms preclude our ability to make those active, and often sensible if subtle connections and relationships between what appear to be discrete occurrences in, for instance, "economics," "foreign affairs," and "the environment." To be sure, an argument could be made that the status of Walmart stock is intrinsically linked to our nation's national security.[12] Along with the marketing mechanisms that allow users to recommend pictures, videos and news items to their friends, this illusion of interactivity is one of the key metaphysical manifestations (and economic pillars) of our Digination.

Closely associated with this pseudo-interactivity is the threat of homogenization. With the mainstream news agencies offering the "best" (meaning most technically sophisticated) news, and given the points already made concerning both producers' and users' valuation of content, we may find that the Internet fosters a hidden homogeneity. In his book *Interface Culture*, Steven Johnson described early on how the Internet would become populated by what he calls "intelligent agents" (1997, p. 188). In the next chapter I look closely at the underlying procedural logics of several of the most popular search engines today, but we can already admit a few hard truths. Software that anticipates our search needs, or *push* technology has not as yet delivered on its promise. Earlier manifestations of push technology did not end up proving itself efficient enough in the realm of Web marketing where it was originally tested. Yet the feeling most of my respondents have expressed—that they are getting access to a great deal of information in a very short amount of time—holds true. This of course prompts us to recall the distinction made earlier between information and knowledge. Despite popular opinion on the matter, they are not equal.

Johnson (1997) recognized early on the positive potential the Internet represents. He envisioned a structure of news and information that would be regularly created by news corporations and less centralized news services alike. This would be akin to the way musical tastes evolve: "from the traditional/top-down tendency, that is a one-to-many pattern of corporate and mainstream content, to the more *mature* bottom-up tendency that results in thorough knowledge of the anti-establishment, the subcultural, the independent source, the many-to-many" (pp. 200–204). But Johnson's vision may be seem little more than idle fancy today given the premium users seem to place on time pressures. Certainly the process Johnson described would require people to tap into many of those text-only sites and newsgroups or dig much deeper into the mainstream sites, both activities necessitating a considerable amount of reading.

Unfortunately, Johnson's top-down story more accurately describes so many Internet news user's experience today. While he was making a comparison to the way people's tastes in music change, he does an uncanny job of describing how, when people bookmark or set up personal accounts on mainstream news sites like CNN, Yahoo, MSNBC, USAToday, etc., they tap into top-down systems that dole the "same old favorites, dictating that everyone attend to a smaller, more predictable repertoire" (1997, p. 204). I found the very same trend in my respondents' consumption of blog content. And what Johnson calls "predictable repertoire," others have called "received truths," and "conventional truths" (Herman and Chomsky, 1988); "Half-Truths" (Chomsky, 2001); and "shriveled and absurd" content (Postman, 1985). The connection we can draw between Johnson's musical metaphor and the mainstream news site is notable. Less diversity and a homophilic tendency is a clear result. Again, this all runs counter to the structural possibilities inherent to the Web. And so we can no longer talk only of theory. The problem is one of practice. Useful research and the comparison of news sources requires work (work people don't think they need to invest) and time (which people claim not to have).

SOME PHILOSOPHICAL IMPLICATIONS

The progressive move from the word to the image that McLuhan's brand of technological determinism entails, and the notion of hyper-utilization intermesh with market logic and consumer behaviors and preferences to account for the trends described in this chapter. The entire process, it would seem, is now largely out of rational control. It is a self-prodding system. Web designers are compelled to keep pace with the technical capacities of the technology. This is why we will continue to see the Internet pushing the limits. We are moving toward a full sensory-immersive environment that centers upon image and sound that is very likely due in large measure to an implicit agreement that was in the making decades ago between Bill Gates of Microsoft and Andy Grove of Intel Corporation.

So if one accepts the argument concerning the difference between words and images put forward here, then it is reasonable to think that, as the technology progresses (with progression being determined almost exclusively by bandwidth capacity today), the quality of content of Internet news sites will degrade proportionately—along with the *content* of users' heads. I see an analogy in the CGI "revolution" in film. Has anyone noticed a dearth of decent plots lines lately? Enough with the morphing figures and massive explosions already! What has happened to our storm-telling abilities?

We should also consider the prospect of interactivity in the context of human isolation which is also, increasingly, a central feature of Digination that began before digitization. It is the tendency, and ability, to make things discrete, to separate and differentiate. As far as Plato, Innis, and McLuhan were concerned, the exteriorization of knowledge brought on by writing (and later extended by the printing press) was, in effect, decontextualizing, alienating, and anti-human. If we are the *body-thinking* creatures these theorists seem to imply, then all extant communication media, by definition, separate us from that embodiment and direct experience. Advertisements over the past decade and a half for everything from personal computers, personal digital assistants, and personalized news services typify the popular trend toward individuation today.

As noted earlier, all communication media/technologies are alleged to carry demand characteristics (specific logics and grammars) which exert causal forces that can impact and change the perceptions, attitudes, beliefs and behaviors of individuals and, by extension, entire populations. Ong (1982) represents an elaboration of the work of Eisenstein (1979), Innis (1951), McLuhan (1962), and others to consider the determining influences previously associated with text, into the realm of modern communication technology. He points out that "[t]he sequential processing and spatializing of the word, initiated by writing and raised to a new order of intensity by print, is further intensified by the computer, which maximizes commitment of the word to space and to (electronic) local motion and optimizes analytic sequentiality by making it virtually instantaneous (1982; p. 136).

Ong was writing before the graphical interface was invented (or at least went public), and he mentions the computer here only in relation to its representation of the word. His point concerning decontextualization gains even more relevance when we consider the swiftly accelerating proliferation of graphically interfaced (i.e., iconic or image-based) electronic media, and the manner in which that proliferation is predicated on the logic of providing individual users with singular control of their machines. In short, as more and more people get hooked up and online, more move into the kind of individualistic, decontextualized perceptual and experiential mode Ong describes. Put another way, the computer renders nearly meaningless the constraints of time and space such that individual users of electronic media today are often *all alone together* and, as Joshua Meyrowitz explains, "everywhere at once, but no place in particular" (1985).

Consider too the increasing presence of CGI animation on many of the mainstream Internet news sites. Taking Ong and Meyrowitz's lead, we might say for instance that *War* is relatively easy to separate from its context. The contention here is that this may be the case because war has become so easy to *IMAGEine*. Peter Jennings, the recently deceased long-time ABC news anchor, commented during U.S. military operations over Kosovo on the fact that with the Internet he could now go to CNN.com any time he felt the urge, and watch a computer-simulated model of a cruise missile firing. The problem, explained Jennings, was that computer-generated cruise missiles always hit their targets. Perhaps Jennings maligned CNN for misrepresenting the truth of the matter because he was feeling the squeeze of lost ratings. Perhaps he didn't like computers. Whatever the case, his observation is also an important insight to remember as we venture headlong onto the digital domain.

Of course we are not talking about getting down to some Platonic Truth. Objectivity in news is a myth surely. The emancipatory trick with Internet news consists in getting people to think for themselves—and this should not mean isolating them. But we have to admit that people do not share the keyboard or touchpad in the same way they once shared the sofa or dinner table. The personal computer, more often than not, remains just that: a private device. Unfortunately, hypnosis is always easier when one is alone with the magician. Like Plato's prisoners,[14] who cannot see others even though they are sitting abreast, the Internet news seeker rarely interacts with others *during the surf*. Despite the air of interactivity hovering around the net, it is still a decidedly top-down and static sort of place for the vast majority of folks jacked in. I can only speculate here, but psychological isolation may be particularly dangerous when it comes to making sense of news increasingly relayed in pictographic form.

C. A. Bowers (2000) considers in detail the problem of isolation and decontextualization. "There is an emphasis," says Bowers, "on diagnosing problems and framing solutions as models that can be replicated in various cultural contexts" (p. 75). This statement, along with Jennings' comments about fantastic content online suggests that we have now entered a time when news can be what, when, and wherever we would like it to be. While talking specifically about computer technology's impact on ecology, Bowers' point that such decontextualized, reductionist "forms of knowledge are inadequate to rectify the problem of moral blindness" . . . and "ignorance of the *larger circuit*" (ibid) holds sway in other areas (including discussions of poverty, hunger, corporate malfeasance, and ecological degradation). Likewise, Jennings' remarks concerning US weapon systems in action demonstrated how computer image processing can practically nullify the real-world significance of war.

ARE WE GETTING THE PICTURE?

While most communication/media scholars have positively rejected the practical applicability of Shannon and Weaver's transmission model of communication, the general public seems far behind. This is despite the fact that fairly sophisticated digital image processing technology is now finding its way into the consumer's hand (at the time of this writing $100 could buy a 10–12 megapixel camera with all attendant processing software delivering impressive results). So many people still equate the photographic image with reality today—as they did around the year 1900. Have we become hard-wired in this way? Whatever the answer to that question, now more than ever, with 3-D imaging, streaming video, and CGI all coming into their own on Internet news sites, the notion that communication media are neutral tools or vehicles that we use to transfer information should be dispensed with.

Finally, what are some of the potential implications these trends hold for the future of a global citizenry? McLuhan at times held high hopes for the *Global Village*. If France, England, Germany, Poland, and the other nations making up the European Union have maintained somewhat higher journalistic standards to date, the prediction here is that the European Union moves away from the biting satirical news programs once found on the BBC, *Canal+*, and *France 2* (like the always provocative *Campus*). Will the move to the image that seems genetically built into Internet-based news at this point spawn a new generation of whimsical, unreflective, easily swayed individuals? Will we move slowly back to a kind of oral culture with the hypnotic effect of the "village elder" dictating meaning and truth? Will the secondary orality that began with the telegraph come into its own with the Internet image gallery and the power of first impression? Or, perhaps, could all this imagery snap us out of our collective reverie and create a population of cynical, skeptical, and careful assessors of news and information? Even in the field of media ecology, there are folks sitting at each end of this divide.

There is no question that the medium-theoretic perspective tends to ignore many of the novel and creative ways in which users employ their medium of choice toward emancipatory ends. Many studies of electronic media use spanning the last three decades have ignored or overlooked the positive practical implications these employments can hold for users.[15] On that note, it should be pointed out that this is not a question about high or low culture. The Internet long ago transcended any such cultural designations. Instead of making any sort of value claim, I want to ask one more set of questions before closing this chapter. What are the usage trends and patterns were now seeing online pointing toward? Are we seeing the broad dissemination of the kind of news and information that allows for the generation of knowledge? Do we see things that are usable in creating and maintaining a democratic situation and an open marketplace of ideas? These are idealistic questions these days I know. Is it naïve to suggest that the news "is supposed to give us information in order to function more effectively in a complex world?" (Mitroff and Beniss, 1989; p. 10).

Bolstering perhaps that somewhat jaundiced view of the media professor quoted at the beginning of this chapter, I do not cherry pick the data when I say that many young people now consider the self-referential meta-commentary of *Entertainment Weekly* or *US* magazine to be the important, pressing news of the day. But then, while cultural critics submit that there is no accounting for the public's taste, the story may not be a simple matter of personal choice. Clearly, multiple variables are in play. And these are powerful forces at work. It is a complicated synergy of factors, including (1) the news industry's commercial imperatives toward concision and "consummability;" (2) the media deterministic logic associated with the move from the word to the image, the associated tendency away from systematic, sequential, rational thought; and (3) user's preferences: the continued American/Western-Industrial turn toward the valuation of entertainment, speed, quantity, convenience, efficiency, ease of use, etc. No doubt all three factors are contributing to the process of hyper-utilization, and the building of our Digination.

We have media determinism, channel bias, political spin, economic incentive, and the fallibility of human perception . . . the extent and ultimate consequences of this alchemic mix, however, remains to be seen. It is a mire for sure, but a mire we must wade through if we have any hope of holding onto a clear and more grounded understanding of what it means to be human in the twenty-first century. As far as a final prognosis is concerned, we may progressively find that as more and more of us go online to gather our news (or have it pushed toward us through any number of digital means) we can't help but fall headlong back into Plato's cave—a place full of isolated ignoramuses making much ado about nothing in particular.

NOTES

1. This is occurring through agents and institutions as seemingly disparate as the American corporation (like Intel and Microsoft) and the American teenager, with some fairly recent teenagers now even running several of the most powerful American media corporations (such as Mr. Zuckerberg of Facebook and Mr. Page of Google).

2. The first "K Quiz" I administered was to students in my mass media and speech courses in the Spring of 2000. While names like Tony Blair and Jacques Chirac and places like the Sea of Tranquility were included in that older version, more recent iterations of the K Quiz include items like: "the name of the vice president," "what and where is the District of Columbia?" "why bloggers in places like Iran and Libya were getting a lot of attention?" and "what happened in Haiti? And Japan?"

3. We now live in one of the lowest moments of political participation, with between 50 and 60% of registered voters actually going to the polls. Numerous explanations have been proffered to account for this state of affairs, including the economic, cultural, and technological.

4. The term refers to a sort of mania about, over-employment of, and/or preoccupation with, some device or idea. Hyper-utilization implies that there are sensible ways to employ the tools and ideas we have developed, but also that the calls to consider any range of more reasonable, modest and restrained applications of these things go largely unheard. Americans unquestionably (and often unquestioningly) engage in the hyper-utilization of automobiles, credit cards, and food. In the last decade we've witnessed and been subjected to the hyper-utilization of the human genome. The last century tells a story of how we have hyper-utilized the image, and we are certainly on course to hyper-utilize the Internet as an information repository and exchange mechanism. But in the case of the Internet, we have to add an important criterion to the definition of hyper-utilization: the notion of the *sensible*. We must take seriously the term "virtual reality." The Internet is a place where we are supposed to be able to participate in activities *as if* we were participating in them in the *real* world. For instance, I recently engaged in what I take to be one sensible utilization of the Internet. I bought a plane ticket to New Orleans. This worked pretty much the same way it would have had I called or walked down to my local travel agent (some wider-ranging economic consequences probably notwithstanding). We might say that my use of the Net in this case functionally simulated the standard activity. It worked in the cyberworld "as if" I did it in the *real world*. I simulated the event and achieved the same real-world ends. But we know that something like the simulation of the firing of a weapon on an Internet news site is not an accurate simulation in this way. It does not represent things "as if." And in most cases, thankfully so. The moral? We probably shouldn't use the Internet (or the image for that matter) for everything—and that list should certainly include the creation and consumption of news.

5. This operating logic, with the heavy premium it places on immediacy and "the scoop," seems to be largely responsible for numerous contemporary reporting problems, including the confusion surrounding the 2000 presidential election and the Florida voting miscalls, coverage of the ramp-up to war in Iraq, the Rather/Bush debacle, the global climate debate, and the numerous hacking and censorship scandals of late.

6. These qualitative changes are by no means limited to the realm of news. Advertising has moved sequentially over the past 100 years from a focus on the particularities of the product for sale, to the vague selling of a lifestyle that is being associated with the product via imagery (the happy crowd inside the automobile, the whimsical walk through the park with Pepsi in hand, etc.). Politics has undergone a

paradigm shift in its modus operandi. Media commentators, perhaps beginning with McLuhan, often point to the first televised presidential debate between Kennedy and Nixon in 1960, as the official beginning of the end of policy-based political campaigning. Henceforth, campaigns move progressively away from content dealing with policy, to an infatuation with the personality of candidates. The argument here is that it is the image that most directly accounts for this qualitative shift, with the visual-based coverage of politics playing the dual roles of initial prompt and continued promulgator of irrelevance and triviality.

7. It is interesting to note that the introduction of Microsoft's *PowerPoint* and other visual presentation software has, according to a number of observers, probably accelerated the expectation of concision. Nadine Dolby (2000) has commented upon the way in which the oft-used templates offered in the *PowerPoint* program tend to "marginalize critical discourse that depends on engagement with language and ideas," and that "the educational profession needs to explore ways to preserve the importance of the word at a time when it is under growing threat." Of course, the threat extends outside of academe. "By their nature," continues Dolby, "presentations structured around overheads must lean towards summaries, bullet points, graphs, charts, key ideas, and other such truncated written, visual, and verbal expression" (p. 1).

8. Godfrey Hodgson (2000) associates this trend with the end of narratives (what he calls "Grand Narratives") that at least try to make sensible connections between otherwise discrete news items. The consequence, says Hodgson, is a growing disinterest in news.

9. See MacDougall (2001, p. 253) for a partial account of the lauding that has accumulated around the Internet.

10. While the people, places and things included in the "knowledge" quiz administered to respondents after they returned the six-item questionnaire is really just a lot of discrete nouns, the paltry scores indicate that many individuals possess a frightful lack of understanding concerning not only the brute definitions and descriptions, but also the relationships and significances of things that in fact relate centrally to their positions as students, citizens, voters, etc.

11. References to *The Onion, NPR, Reuters* and other news sites that typically feature more lengthy, text-based news stories, came low on all of my respondent's lists (it is conceivable, however, that a larger, more random sampling may not align perfectly with the current finding).

12. Consider also the very real and intimate relationships that exist between, say, automobile sales, oil production, gasoline prices, foreign aid, famine, poverty, terrorism, and war.

13. The note of cynicism in this last statement is intentional. The discontent surrounding the constantly updated *Windows* operating system is, at this point, a kind of cultural lore.

14. The reference here is to the "Allegory of the Cave," a dialogue designed to generate reflection upon the nature of knowledge and education found in Plato's *Republic*.

15. For the negative account see, in particular, Culnan and Markus (1987); DeSanctis and Gallupe (1987), Kiesler, Siegel, and McGuire (1984), Sproull and Kiesler (1986). For a collection of more charitable stories see Bench-Capon and McEnery (1989), Boczkowski (1999), Coats and Vlaeminke (1987), Feenberg (1993), Furlong (1989), Guldner and Swensen (1995), Jackson (1996), Kerr and Hiltz (1982), Lea, O'Shea and Fung (1995), Lea and Spears (1995), Meyers (1987), Rice (1987), Rice and Love (1987), and Walther (1992, 1993, 1994, and 1996).

Chapter 6

Search Engineering and Our Emerging Information Ecology

"Not everything that counts can be counted, and not everything that can be counted counts."

—A. Einstein

Extending the discussion in the previous chapter, let's now lift a corner of the veil that occludes an increasing percentage of our information gathering practices today. Artificial intelligence systems underpin so much of what we do when we do things online. Consider news gathering, or just *getting the news*. The two most popular news and information aggregation services today are *Google News* and the news-dedicated arm of the *Yahoo* search engine/portal. In terms of market share *Google* dwarfs *Yahoo,* and Microsoft's *Bing* has been a recent upstart in a concerted effort to siphon off some of the network effect that has been working in *Google's* favor for more than a decade now. Indeed, if our Digination were a living, breathing organism, *Google* might be its heart and lungs. While Microsoft's late entry in the browser wars has some features that suggest it to be somewhat unique. *Bing* however came on the scene well after I collected the data for this chapter. I undertake a comparison of the primary data-mining strategies employed by *Google* and *Yahoo* to reveal different proportions of human and machine-driven strategies of information management that remain more or less state-of-the-art today. Much can be extrapolated from this comparison to offer some insight into the types of information structures being generated for public use today.

A basic understanding of the computational logic of these two systems helps explain some of the systematic difficulties users may confront when trying to discern authoritative and more accurate information, from the more speculative, misleading, even fallacious. We'll discover that the channel/source distinction is one of the central conundrums of our Digination, so the comparative analysis is followed by a critique of the increasingly opaque news and information environments now emerging in and through them. Finally, I'll note several problems that may contribute to an erosion of the social and political health of any society connected to such apparatuses.

Given the small scale of this study, a conclusive report on the effectiveness of these two reigning approaches to information management is not possible. However, we can critique trends in news search result offerings at a time when these aggregation mechanisms are being perceived as news sources in and of themselves. We can also highlight some potential side effects associated with the personal, social, and political utility of these systems as news conduits. So my analysis here responds as well to a trend in communication and social studies of technology to bracket off the non-human or the ostensibly "non-social" components of systems. As described early on by Mumford (1934), Innis (1952), McLuhan (1964), Carey (1989), Latour (1992), and others, we begin to see how machines are left out of the intentionality/action equation at our own peril. This notion is wonderfully captured in James Carey's (1989) suggestion that "machines have teleological insight." Or, to put it another way, any machine, technology, or medium has certain ends *in mind.* Without anthropomorphizing this idea beyond intelligibility, Carey's point should be well-taken. There's no doubt, search-engine technologies become an active participant in our everyday communicative and sense-making processes, practices and outcomes. The upshot is that we may be witnessing a progressive dilution of intention and agency in humans with respect to whatever mastery we might claim over our information environments.

AOL, Yahoo, and *Microsoft* continue to grapple for position as *Google* becomes the default news and information channel for the majority of the US population.[1] If we do appear to be reaching a state of corporate equilibrium online, making sense of the profusion of information available there remains a daunting task. A full-blown analysis of the search logic underlying *Google* alone would certainly help media researchers understand a variety of important communication patterns and processes now emerging. I offer some first steps toward that in this chapter by focusing on the way different search engine logics alter the macro structures of information available to Internet users and, subsequently, the individual and collective structures of knowledge produced and maintained by them. From this we can begin to see how common user search strategies, conventional understandings of the relevance of search results, and the interpretations and emerging meanings of "authority," "expert," and even "truth" might contribute to the construction and maintenance of communities *on-* and *off-*line.

The underlying mechanisms of these two leading information assessment, association, and retrieval tools suggest patterns of knowledge creation that are beginning to resemble something like the central arbiter of a public mind. Understanding our Digination requires that we learn something about the way this mind works. I know this: *Google's* current instantiation is probably not what Marshall McLuhan himself had in mind in the late 1960s with his notion of a *Global Village* (McLuhan and Fiore, 1967). Primarily non-human search mechanisms also seem in many respects antithetical to what Tim Berners-Lee envisioned would emerge from the universal server/client software he wrote in 1989 that has become the lingua franca of the World Wide Web.[2]

The fundamentals of search-engine technology are relatively simple to understand if we consider two basic theories of information assessment and retrieval. In different ratios, these approaches combine with conventional information retrieval methods to result in virtually all of the mainstream search engines now running online: the fully automated, beginning-to-end artificial intelligence (AI) approach to item indexing, referencing, and for-warding, and the approach that incorporates human editors and indexers in the initial information evaluation stage. These are the general approaches of *Google* and *Yahoo* respectively, so I limit the discussion in this chapter to these two reigning browser architectures since they are currently the front runners in a race toward what may end up becoming an industry standard.[3]

The condensation now occurring in the browser wars is in keeping with long-held patterns of technological innovation (Rogers, 1995). Once a single format begins to gain momentum via commercial marketing and PR, word-of-mouth promotion, and brute installation, the wave of adoption and diffusion is often nothing less than remarkable. *Google's* rise over the past decade has indeed been extraordinary. It is the fastest growing company in the history of telecommunications. General public acceptance of what is perceived to be the leading technology or product is a pattern noted in the *diffusion of innovation* literature (ibid). This arbitrary aspect to so much of what we take for granted in the world around us often contributes to the staggering growth of just about any technology that catches on, and the cases of *Microsoft, Yahoo, AOL,* and *Google* all apply. "To *Google*" and "*Googling*" are fast becoming stand-in terms for the general activity of searching the Web, and so many alternative search methods are systematically overlooked, or possibly never even heard of. *Teoma* (running under *Ask.com* at the time of this writing) for instance, is a very efficient and search-specific engine that has been given short shrift due to the near-total domination by *Google, AOL,* and *Yahoo. Google* is fast approaching the eponymous status of *Kleenex* or *Coke*. Bookmarking *Google* or *Yahoo,* or making *MSNBC.com* the home page on one's personal computer creates a default Web portal and further contributes to a powerful *network effect* (cf. Ferguson, 2005).

PREVIOUS RESEARCH AND SOME
BACKGROUND TO THE PROBLEM

If one doesn't like the heart-and-lung reference above, information portals like *Google* and *Yahoo* are also access points to the Internet, the spine of our new information ecology. These systems put constraints on how human beings parse, organize, and make sense of the world around them. Patterns and processes of information assessment and control have long been a central concern of network theorists, information scientists, librarians, and educators. Communication scholars interested in the way mass media audiences process information have also developed a number of theories to understand how people make sense of their worlds (Sundar and Nass, 2001; Wicks, 1992). AI applications that help humans parse information online have altered these sense-making processes. The public is now offered pre-made knowledge structures in the form of search results listed in rank order that have been generated in many systems by the logic of a proprietary algorithm that uses popular taste as a primary discriminator. Given this, communication scholars can no longer ignore the personal and political significance of commercial search-engine technology being deployed on the Internet today. There is a solid tradition of research in this general area of mediated sense-making.

Stephen Chaffee (1982) elaborated on the practical and conceptual difficulties that arise when seeking to distinguish between *source* and *channel* in mediated situations. If portions of Chaffee's analysis are now clearly dated, several core problems he noted have been exacerbated as the Internet stands ready to usurp television as the primary medium in the United States. Chaffee's original observation still holds. He suggested that media consumers' channel and source preferences tend to be informed by two variables: (1) *ease of access* and (2) *the likelihood that the source will contain the desired information.* Counter-intuitively perhaps, the source/channel distinction becomes even murkier online, as problems of channel accessibility and the likelihood of finding desired information appear to have been largely alleviated.

Chaffee's two heuristics retain their function for many media consumers seeking information on the Internet today. However, the intervention of AI-based parsing mechanisms significantly blurs the meaning and import of the "desirability" variable. Depending on the topic of the inquiry (from news about Iraq, to maintenance directions for a specific make and model of car, to local opinions about seafood restaurants), information seekers often do not have a clear sense of what they desire, at what point they might have found what they were looking for, or if the information they settle on is even reliable. *Satisficing* and *bounded rationality* are terms that have been used to describe the character of some mainstream Internet inquiries today (Agosto, 2002; Resnick and Lergier, 2002; Wallis and Steptoe, 2006).

I know from talking with my students over the years, as the Internet and its management tools have matured and taken shape, that many folks still have a lot of trouble with these distinctions Chaffee mentions. Consider the whole "wiki" phenomenon. These are wonderful tools if used and understood properly. Unfortunately, the number of folks who fundamentally misunderstand the nature of *Wikipedia* is astounding. A majority of my own students still enter their freshman year with the impression that *Wikipedia* is an information source when it is, in fact, more akin to an information repository or clearinghouse. We clearly have a long way to go as far as getting our population up to speed with even a basic media and information literacy is concerned.

Many Internet users will also readily admit to having trouble determining the originating source of the information obtained when using *Google, Yahoo,* or search engines. Ironically, the awareness of abundance or scale also tends to foster a general sense of reliability. So an attendant issue arises in the realm of Internet news consumption: the underdetermined quality of information made available to the public.

If access to a broad array of news and media outlets online has become a matter of typing short search phrases and then choosing from a set of results, it is not obvious to me that we now have a situation in which Internet users are finding the information they *desire*; if by this we mean to describe some intentional stance, or the successful satisfaction of a personal information need.

Here again, I'll speak to my primary source of experiences in this matter (professional educators) and say that many of my colleagues agree that there is a lot of tenuous happenstance going on these days, with a sort of "first-must-be-the-best-and most-relevant" logic at play in many college-student Web searches. The reference pages of term papers and final project betray the reality from community colleges to the Ivy League. I can't speak with as much accuracy to the public trend, but I also get the sense that it is not entirely unlike the student sample. And this is a very different sort of happenstance that occurs in the traditional search-and-stumble-upon effect that is endemic to a visit to the library stacks. In that case, it is the topical and thematic organization of information that guides the search. I probably would have not discovered McLuhan's intellectual mentors Harold Innis or Lewis Mumford until much later had it not been for a tinge of ADD manifesting itself during my information safaris into the massive collection of communication and media studies material at my old school.

Indeed, while information literacy courses and workshops are becoming more common in higher educational contexts around the country, there is still much that needs to be done as far as getting the coming generations of citizens, leaders, decision-makers and thinkers up to speed. If such courses and workshops are often designed well enough and are very often even required, more often than not they are conducted in ways that don't really ensure students get the material—that is, acquire the requisite awareness and skills necessary to separate the wheat from the chaff. As my favorite hipster librarian put it during one such workshop that I required in a fundamentals of communication course: "[l]ook, the Internet is really really big, there's a lot of information out there. But remember, about 93 percent of it is garbage." Now while that number was uttered off-the-cuff, and is probably disputable along several lines of attack, the estimation is valid in more than just a rhetorical sense, as I'll point out shortly.

If Chaffee was writing about the traditional mass media outlets of his time (i.e., print, radio, and TV in the late 1970s), his insights concerning the increased difficulty media users experience in distinguishing between source and channel have gained even more importance as the population moves to online venues for their news and information. If the Internet has yet to deliver on its promise of enhanced knowledge and information for the betterment of all, this remains the "democratic promise" vended by noted public intellectuals, industry leaders, and politicians including Bill Gates, Steve Jobs, Sergey Brin, Larry Page, Nicholas Negroponte, Yochai Benkler, William Mitchell, Al Gore, Bill Clinton, George Bush Jr., and Barack Obama. It is an idea encapsulated in the claim: *the more [information], the better*.[4] This is, unfortunately, an idea that has spread along with the ongoing process of Digination. It becomes, in short, a habit of mind. It may, however, be a debilitating habit that short-circuits critical thinking and stymies our efforts to stop gathering and periodically look under the surface of the lists of files, images, and documents saved to now tera-scaled memory allocations. Nonetheless, an AT&T commercial on television recently asserts that *"Access to the Internet makes people smarter."* The text superimposed over what appears to be a middle-school spelling bee gone long continues: *"What will happen when everyone has access to it? Let's See."*

This tacit assumption of the digital age—that we just need more data—suggests that if we increase the amount of information available to the public, the public will then find what is required to better fulfill the demands of citizenship. Other prognosticators spanning the twentieth century, including Aldous Huxley, Walter Lippmann, Marshall McLuhan, Tom Patterson, Neil Postman, and C.A. Bowers have argued to the contrary. Most generally, they contend that this age of information abundance will or has resulted in a confused citizenry ill fit to process the many nuances of our socio-political world. Indeed, there is good reason for confusion. As I'll illustrate below, the

Internet is an information environment designed to be parsed and processed by artificial, not human intelligences.

METHODOLOGY

This book is organized around a one-medium-per-chapter format. Of all the media analyses found here, the present chapter is probably the most social-scientific of the mix. Whereas most of the other chapters incorporate ethnographic components to bolster the phenomenological perspective I'm promoting over all, the most qualitative aspects of this search-engine analysis are the observations provided by me, a motley crew of students and colleagues, as well as the ruminations of two graduate research assistants. In this respect, it should qualify as a participant observation analysis of Web search logics. At least that's how I went about obtaining my "raw data." I then get more speculative in my interpretation of the data toward the end of the chapter.

All of the data gathered for this analysis were obtained from the *Google* and *Yahoo* search engine portals at regular intervals for ten consecutive weeks during the Fall and Winter of 2005 resulting in 100 hours of search activity, with about a quarter of that time devoted to a third search engine.

A graduate assistant studying integrated marketing agreed to surf the Web using "mirrored" searches six hours a week. The mirrored search required the input of verbatim (copied and pasted) search terms and/or phrases into the *Google, Yahoo*, and later *Teoma* dialogue fields. Directed topics and search language were based upon items in the political news headlines at the time, including the war in Iraq, the personal details of two Supreme Court nominees, and the immigration debate. I also provided a list of basic search terms including: "troop morale in Iraq," "upheaval in the Bush administration," and "Alito's bid."

In an effort to obtain as naturalistic an impression as possible of the similarities and differences between the search engines, directed inquiries were limited to 2 or 3 hours each week. Additional search topics and terms were incorporated ad hoc by the research assistant during ongoing work on school projects or for personal interests. Some of those search terms included: "Walmart going green," "viral marketing" and "social marketing," "spam and social marketing," and "James Bond and product placement." All searches were mirrored during work sessions and saved to file. My assistant developed an efficient copy-and-paste procedure with multiple windows open to the respective search engines that streamlined the procedure to an extent, which I then adopted during my own mirrored searches. Nonetheless, the routine was cumbersome and time-consuming, with an additional requirement to always delve at least two layers into search results: the *list* and *content* levels. The list level was defined as the actual list of search result items. This is a typical search result item on *Google*:

> Troop Morale in Iraq—New York Times
> Troop Morale in Iraq . . . For those soldiers interviewed who reported high morale and an understanding of their mission, it would be extremely informative to . . .
> query.nytimes.com/gst/fullpage.html?res=9D00E7D91731F934A15752C1A96
> 39C8B63—18k—Cached—Similar pages

The *content level* is the level of the news item content or document itself. Using the *printscreen* image capture function, the first ("listing") and second ("content") levels were catalogued sequentially. To obtain sufficient sourcing data, up to 200 items were catalogued for each search. However, 20 to 50 items were typically enough to demonstrate a significant difference in terms of quantity and quality (or breadth and depth) of sourcing that was tracked between the *Google* and *Yahoo* approaches to information assessment and retrieval.

DEMANDS FOR SPEED, EFFICIENCY, AND THE QUALITY QUESTION

In *Darwin Among the Machines* (1997), George Dyson suggests that the "diffuse intelligence of the web and all the nodes of processors has more in common with Hobbes' leviathan than with the localized artificial intelligence that has now been promised for fifty years" (p.7). While Dyson was referring primarily to small servers and personal computers functioning as nodes in a larger system, his analysis of modern computer networks eventually extends to include human users working symbiotically with their machines to produce information structures unprecedented in scope and scale. Efficiency (in terms of speed and cost), and quality (writing and research) have been central concerns in the fields of computer science, information and library science, and journalism as long as these have been viable professions.

Tracking the creation, collection, and consumption patterns of news is a tedious undertaking, butitisa good way to see the relationships between efficiency and quality of information online. Beginning with the regular use of the telegraph by newspapers near the end of the nineteenth century as news organizations began to depend more heavily on advertising to support their operations in the early twentieth, the complexity of electronic communication systems have progressed steadily. Efficiency and quality have also increasingly been at odds. In the 1980s, as *CNN, Reuters,* the *Associated Press,* and other 24-hour-news services gained solid footholds with foreign bureaus peppered across the globe, the public call for immediacy, and *all news all the time* became a realistic expectation of news audiences and, eventually, an institutional dictate for the larger news agencies themselves.

News itself becomes a *channel*, not just "the news." The 24-hour-news concept and the *CNN effect* create the need to fill time. In the process, all kinds of information, including raw data, statistics, and the listings of disparate events begin to be treated as news. The term *infotainment* is coined in the mid-90s and Matt Drudge, Perez Hilton, and Arianna Huffington become de facto news institutions. As large and small news outlets continue to steadily populate the digital domain, these expectations become the order of the day. The beta version of *GoogleNews* was launched in September, 2002, and by the end of 2006 allowed users free access to thousands of news stories from around the world. The system automatically selects and compiles news headlines and images, and updates its pages every fifteen seconds based upon transformations in the Web's interlinked structure.

It is generally acknowledged that *accuracy* and *depth* are descriptors of *quality* concerning news content. Given this, we see how a blurring of the quantity-quality distinction emerges as a predictable by-product of any information reservoir like the Internet that, practically speaking, has no memory or processing limitations. Unlike the rows and stacks of a traditional library, the very finite "news holes" created by the spatial constraints of commercial newspapers and magazines, or the temporal constraints of radio and television, there is no news or information hole online, so to speak. Given this, it is arguable that the massive amount of information available on the Internet today is not suited to effective parsing nor practicable comprehension by human minds at all. This is a limitation of our own design as human beings. Ours is a *bio-social design* that includes embodied, localized but extended minds with limited and fallible memories. By contrast, AI programs have been designed precisely to deal with this unprecedented scale and complexity in efficient ways.

If numerous factors account for the current confusion experienced by Internet users regarding the discernment of a message's source and channel, the blurring and overlapping references to *media producer* and *media consumer* has also contributed to the puzzlement. And there is still another reason for the confusion. It concerns the sheer vastness of the *information ecology* we are creating, and it is an obvious point to make. As the overall amount of information on a given topic increases (be it Barack Obama's questionable investments, or Scooter Libby's fall from grace), the quality of the information is simultaneously enhanced and degraded depending on where and how one looks. While there is generally an increase in the amount of reputable and accurate content online, there is a concomitant though disproportionate increase in the amount of disreputable and inaccurate content posted on websites, blogs, and news threads. For the end user then, as quantity increases, news quality tends to lag further and further behind. In some respect, then, the Internet is a case study in the law of diminishing returns.

Of course, the public's recognition of truth and accuracy has always been situational and often subject to coincidence and even luck. For instance, Walter Cronkite came to be "the most trusted man in America," the gold standard in tele-journalism, due to what really ends up being a series of socio-historical contingencies. Above all, perhaps, is the fact that television was beginning to attain the status of contending news medium throughout Cronkite's tenure as anchor at CBS (1962–1981). This, along with Cronkite's being in the right place and having the right looks and sound for the time certainly contributed to his status as high priest of political news. Without taking anything away from his talent as an anchorman, Cronkite certainly owes something to chance—his particular placement along the arc of modern technological change. However, as the sobering effect of history often reveals, we have come to learn from Cronkite himself that his trademark phrase "And that's the way it is" was not always a statement that could be supported by the facts. Still, as the prime *channel* through which millions of Americans gained their picture of the world, Cronkite also, in effect, became the most trusted news *source*. Of course, Cronkite was no more the source of the news he relayed than Matt Drudge was the source of information about the Clinton-Lewinski scandal. Drudge assumed the status of default source during the scandal because the typical mainstream sources (*AP, Reuters, Bloomberg, CNN,* and *The New York Times*) initially opted out of reporting the story. Drudge, like Cronkite then, was a clearinghouse, a news conduit, a node with more authority than usual, in the information matrix.

Google is also, in part, a champion of circumstance, both historically and technologically, that has blurred the sense/nonsense and source/channel distinctions in similarly serendipitous ways. Individuals and websites can become "source-like" in this age of hyperlinks due to their superior ability to bring together disparate new items that suggest a larger picture of things and which at least *seem* to make sense to large numbers of news consumers. Although the popular blurring of source and channel did not start with the advent of network television news, the common misapprehension did firmly establish itself with the rise of Cronkite and his ilk. Akin to Cronkite's establishing a dominant ethos in the network news business in the 1960s and 1970s, *Google* has very quickly amassed a powerful *persona* of its own.

And yet for some commentators these questions of quality and authority do not seem to be any cause for concern. Nicholas Negroponte's *Being Digital*

(1995), and William Mitchell's *E-topia* (1999), both written fairly early in the life of Internet news services as we know them today are examples of the tenacious optimism that continues to infuse popular scholarship addressing "information flow" and "network" questions in particular. Anthony Wilhelm's *Digital Nation* (2004) keeps the party going. For these writers, the leveling effect of the Web might produce confusion and chaos initially, but the disproportionate power structures of history that sustained them through the twentieth century are bound to crumble in the face of an unstoppable "digital democracy" writ large in the twenty-first. According to these writers, reality, facts, evidence, and ultimately, the truth will percolate to the surface, and it is the unprecedented network of individual minds thinking, working, and questioning in massive parallel fashion that is best able to recognize the truth when they see it.[5] However, in the decade that has lapsed between Negroponte and Wilhelm, we find that none of this has come to pass in any appreciable way.

Injecting a sobering dose of pessimism into the mix, Manuel Castells' trilogy (1996, 1997, 1998)[6] details some of the social, cultural and economic phenomena associated with the dark underbelly and shadowy corners of our digital domains. Castells' one thousand page volume is peppered with positive instances of what he terms "network logic," but he mostly offers a macro-level look at the deleterious effects of the global reach of networked crime, posits new forms and scales of dangerous fundamentalism, and predicts a widening gap between the information haves and have-nots. Castells' notion of "real virtuality" echoes Baudrillard's (1988) description of the *simulacrum*, and reifies the establishment of a new kind of underclass outside the global information economy that typifies developing nations. Though, as much as I love reading Baudrillard, one shortcoming in both of these authors' approaches may be their tendency to over-abstract the concrete social realities emerging from the network processes they critique.[7] Having taken into account some of the core theory and research framing the current study, we are now in a position to consider what it means to be *authoritative* in our Digination today.

ENCODED VALUATIONS OF *AUTHORITY*, *CONTENT*, AND *STRUCTURE* ONLINE

The first thing we have to understand about Internet search engines is that they do not actually search the Internet every time someone inputs a search phrase and presses the enter key. If they did, we would not get 11.6 million results in a "wait" time of as little as 0.20 seconds when we type in the search phrase "Obama approval rating." Instead, what Google does is create a sort of best-guess of this massive web of information and updates it every fifteen minutes or so. The *root* and *base* sets of information stored on thousands of these relatively small servers create a sort of mirror of the information available on the Internet at any given moment in time. *Google, Microsoft,* and *Yahoo* all *operate* massive server farms that each suck more power than your average New England town, and such farms are where all the searching happens. If there are temples representing Digination, these would be them.

The basics of search engine design and function can be grasped by considering the distinction currently being made between "hubs" and "authorities" online, and the three theories or logics of information retrieval and analysis: *Content Mining, Structure Mining* and *User Mining.* Simply put, a webpage that is "pointed to" by many other Web pages via hyperlink is deemed an *authority.* A Web page that points to many other pages is referred to as a *hub.*[8]

As far as the three "mining" strategies are concerned, several subtle differences set apart these reigning search logics. Robinson (2004) suggests that since the Web really is a very disorganized place, "[s]tandard Information Retrieval (IR) techniques alone would rank junk pages highly if they happened to contain multiple instances of the search term. At the same time, the techniques would miss relevant pages that did not include the term at all" (p.2). The *PageRank* algorithm powering *Google,* and the *Kleinberg/HITS* (hypertext induced topic selection) algorithm running under *Yahoo* searches are, in the end, very similar constructs. Both *see* the Web's *structure* they periodically copy to their servers to be of primary importance in determining the rankings of pages. For this same reason, both algorithms become subject to a phenomenon known as "topic drift." One common instance of topic drift can occur when a search term is ambiguous due to the existence of multiple referents in the dataset. For example, I can type "Green Beetle" as a search term in my favorite browser and get back results in reference to the car or that species of insect. In an effort to combat topic drift, competing algorithms employ different methods of search focus and control.

> While structural information provides us with much information about a query, additional information seems necessary. Kleinberg's algorithm [underlying Yahoo's and Teoma's search logic] also uses only the top authority scores, but there may be useful pages that rank strongly as hubs. Since web queries are an application driven towards maximizing user satisfaction, we can use user feedback to try and weight hub and authority scores so that we can classify "better" results (Nilsen, 2003; pp. 27–8; brackets added).

User satisfaction remains an ill-defined concept. Does it imply usability or "user friendliness," or does it point to the degree to which someone—or some group—might be satiated after a search has stalled? The ostensive meaning of the phrase remains unclear. However, in an effort to combat "such drift," *Yahoo* employs *Web Content Mining* in some of their applications to keep their machine on track at the time of this writing. With the help of human editors working for *Yahoo's* news service, and the incorporation of constant user feedback in its calculations, *Yahoo* search results tend to be slightly more focused—both in terms of relevancy and size. While this is not always the case, it is a trend widely reported in the computer science literature, as well as a pattern corroborated while collecting data for this project.

Web Content Mining (WCM) is a context-sensitive, search-specific, and therefore more labor/calculation-intensive method of parsing and organizing information on the Web. What WCM lacks in speed and efficiency it ideally makes up for in relevance and quality of results. In fact, WCM may be the best search logic for news-gathering since, ideally, anytime news is the subject, relevance and quality should be primary concerns. *Yahoo* seems to acknowledge this with an emphasis on WCM. WCM conducts an inner- (or intra-) document/page analysis of its stored data, with an automatic function that combs the static content within this relatively small universe to establish an initial scaling or ranking of documents/files on the Web that are most relevant to the source from which the query was generated. One early example of a system that employed this approach is the *Watson* program developed by Budzik and Hammond in 1999. Watson was an early instantiation of an *Information Management Assistant* (IMA).[9] IMAs are designed to perform intra-document searches, relating words and phrases within the same document while it is being written. In sequence, using a fairly rudimentary kind of semantic analysis, this type of system first combs through terms and phrases within a document under composition and then "intuits" the meanings of the relationships between those terms and phrases. The system is also simultaneously searching the *Web mirror*, and periodically returns suggestions regarding documents posted online that may be of immediate relevance to the author in the composition of the document at hand. Today, IMAs still employ intra-document analysis as an elaboration upon WCM, a key feature of the *Yahoo* search philosophy.

> From the vantage of the human user, a natural language interface appears to be ideal. The user simply enters a search request. Users need only express their request in terms of a few search terms. What could be easier? Unfortunately, even if information retrieval systems attempted to understand and represent the concepts being expressed in the documents they index or the requests they process, they are divorced from critical information that is necessary to understand them. Namely, traditional information systems are isolated from the *context* in which a request occurs (Budzik and Hammond, 1999).

The significance of *Watson* and Web Content Mining to the current discussion of large Web browsers like *Google* and *Yahoo* is two-fold. First, *Google* and *Yahoo* do not perform content mining and analysis in the same manner. These systems also do not compute the same authority weightings obtained from intra-document mining. *Google's* intra-document mining technique is really a form of content analysis that performs keyword-density sweeps of discrete documents, whereas *Yahoo* does this in conjunction with measuring the assessments of in-house editors as well as user feedback statistics regarding the semantic content of those documents. Garrett French, a staff writer for *WebProNews.com*, reviewed a recent study by *GoRank.com*, an online service that conducts analyses of search engine technology, allowing content and service providers with empirical research to help "unlock the secret of ranking 1st in the search engines" (French, 2004; p.1).

Comparing the data between *Yahoo* and *Google*, French points out that "[t]he first major difference that jumps out is the preference *Yahoo's* algorithm seems to have for more words on a page. The average number of words on a page for *Google* was 943, while *Yahoo's* average words per page in the top 10 results was 1305" (ibid). What could account for a 30–35 percent word count differential between *Google* and *Yahoo*? One suggestion is that *Yahoo's HITS* algorithm spends more time ferreting out the relationships between intra-document terms and phrases to get a better sense of their meanings, and by extension, the larger relevance of the document to the user's search query. [10] That is, the larger word count values may be a formal consequence of different timeout settings assigned to each program. But given that timeout settings are proprietary, the point can only be speculative at this time. Particularities of the underlying mechanism notwithstanding, *Yahoo* (in conjunction with direct human inputs), devotes more resources to WCM than *Google*. [11]

The *GoRank* study also reveals that the average keyword density (the number of times a search phrase or item shows up in a particular document) for *Yahoo* was 19.6 percent, where *Google's* keyword density is 16.9 percent for the results compiled (ibid). While the difference is subtle, this may often be enough to account for the wider range of sources generally returned by *Google,* the longer result list itself, and the sometimes questionable relevance when using the identical search terms and phrases. Put simply, *Google* will provide a larger number of documents that contain fewer instances of a given term. All other things being equal then, *Google's PageRank* glosses the intra-documental level of the Web. [12] One industry observer suggests why: "What is different is *Yahoo's* systematic plan to build 'community intelligence' into nearly all aspects of its operation—and in turn, to entice users to spend more and more of their time on *Yahoo* sites, where they can see *Yahoo* ads" (Fallows, 2005; p. 2).

Web Structure Mining (WSM) stands in marked contrast to content mining because of the way it views the structure of links between documents (or files or pages) on the Web to be the best indicator of relevance and authority. WSM is the primary method *Google's* algorithm employs to determine the authoritativeness and relevance of information it returns to its users. WSM is also the logic employed by *Amazon.com* when it alerts someone, upon purchasing a book, that other customers who have purchased that same title have gone on to buy title x, y, and z as well.[13] I find *Amazon* to be on target about half of the time in terms of providing me with new book information that might be useful to me. *Amazon's* automated referral system *assumes* that if others have bought a particular combination of books including the title I purchased, then the content of the referred books must be related and, by extension, relevant to me. Notice that *Amazon* does not verify if anyone has, in fact, *read* any of the books in question. This is because WSM considers the *structural relationship* between nodes to be the operative heuristic, not the semantic content of information within the nodes of a given universe.

Google, and other WSM-type aggregators perform inter-document/page level analyses by weighing the changing "authoritativeness" scores/rankings of uniform resource locator (URL) addresses by sampling the larger hyperlinked patterns between those pages. *GoogleNews,* for instance, mines somewhere between 4,500 and 5,000 news sources in a given search query and updates that source list "continuously" (that is, every fifteen minutes). The source list is essentially a collection of global "news outlets"—some commercially affiliated with *Google*, some not—composed of massive entities like *CNN, Reuters,* and *The New York Times,* to much smaller operations (potentially as small as a lone blogger working out of his or her home or from a mobile device and *on the run*).[14] Based upon the latest inter-document clicking statistics of millions of *Google* users worldwide, the list of 4,500 ebbs and flows accordingly. Both *Yahoo* and *Google* also tap the *blogosphere* to deliver a substantial portion of their news queries. The results obtained in late 2005 for the present study indicate that the *Yahoo news* diet was 31 percent blogs, while *GoogleNews'* inclusion of blogs hovered around 20 percent.

Yahoo's *Kleinberg/HITS* algorithm works from a root set established by human editors' assessments of the content of specific documents. The root set is the underlying or originating collection of web pages from which a system initiates its search cascade. *Yahoo* also incorporates intra-document and community/user data to maintain the base sets (the wider collection of Web pages that statistically emerge from the search) for more "dedicated" functions like *YahooNews.* This helps explain why a news search via *Yahoo* using the identical search term provides a result list with a somewhat smaller number of discrete sources or content providers than *Google. Yahoo, Google, Bing,* whatever qualifies as one's favorite search engine, I'd argue that the search results appearing at the top of the first screen amount, for the vast majority of information seekers today, to an argument concerning what is right and true. It is a persuasive appeal about what's worth thinking about, what is important and what's not. If some hopeful future eventually makes things different, for the time being our online news-scape is still dominated by corporate interests. Given this fact, there is a sort of emergent agenda-setting function built into these systems.

During the data collection phase of this study it was, however, common to see more blogs, and more blogs sitting high in search result lists on *Yahoo.* Because of this, the number of discrete sources becomes very difficult to discern and the actual base set is further obscured. But we have to keep in mind that the clearinghouse nature of blogs also creates a false image of diversity, and a closer look at the content of most news blogs reveals a subtle form of mainstream mirroring and self-referentiality. That is, a disproportionate number of references and links point to established (i.e., more mainstream, corporate) news sources, as well as other blogs engaged in similar collection strategies. In blogs, then, we find what amounts to a human-driven information management model that relies on a variety of WCM, WSM, and WUM combinations and ratios. At the end of 2005 it was in keeping with *Yahoo's* information search and management philosophy to incorporate more blogs into its data universe (root and base sets) than its rival search engines.

If the findings vary, there is a general trend in searches conducted that demonstrates how the base set for *YahooNews* is scaled down relative to the base set referenced by *Google.* These differences are often subtle, and blogs present a new challenge in trying to measure these processes, but the basic observation holds. In the case of *YahooNews,* content from which the wider search, comparison, and analysis is performed is reduced by the intervention of human editors. It is bracketed and framed by their assessments of specific search criteria (including source credibility and currency). If various instantiations of this hybrid human-machine search technology have been in the public domain for nearly two decades now, usable (i.e., more "user friendly"), fully automated natural language processors have only been around for about half that time.

To clarify, in all three cases (Web *Content, Structure* and *User* mining), we are interacting with various mixtures and proportions of human and artificial intelligences. When we interact with AI constructs like the *PageRank* algorithm we are handing a portion of our subjective, qualitative intentionality over to a machine.[15] Of course, as human editors feed an algorithm via WCM, they also do some of the categorization and relevance-checking work for the user. In that case too we are relinquishing a portion of our ability to think and act in an intentional and autonomous fashion relative to some collection of information over to a more coarsely-grained human-machine hybrid. And the same holds for any walk through a traditional library. We are at the whim of the information experts employed there—as well as the very uneven funding allocated to public and private libraries around the country.

With *Google* we simply displace governance and judgment onto a much more ephemeral and fine-grained cybernetic system (multiple machines and millions of users) and what is, via WSM, ostensibly the more objective, quantitative-based measurements and assessments of a heuristic algorithm.

MURKY MEANINGS OF "NEWS" AND "AUTHORITY" ONLINE

Chaffee (1982) pointed out how electronic media were making it increasingly difficult for people to recognize the difference between source and channel in the communication process. Of course, making *that* distinction should be more important than ever in this era of automated information aggregation systems. With the advent of blogs and the meta-search apparatuses described here, "search-specific function" is the buzz phrase Web designers employ in an effort to assuage concerns that the Web has become increasingly confusing as an information repository and unwieldy as a knowledge tool. The phrase suggests that academic searches, news searches, technical searches and Web-wide searches are all distinct enterprises. And this makes good sense in certain instances. For example, if I want to determine the best and most efficient way to repair a faulty idle control valve on my old Volkswagen, I can get that information from a VW-tech blog with a short and sweet search phrase. Similarly, if I am interested in a detailed history of *Google*, I can locate various timelines on the subject. I can even compare *Google's* own history of itself with other views—like *Inc* magazine's story, or *Fast Company* magazine's take on *Google*.

The problem is that once we move beyond these kinds of static data sets, things can and often do get murky—and fast. Reliable information on the inner workings of various search algorithms that I was also able to comprehend was very difficult to locate. Part of the problem stems from the newness of the technology, and the fact that these algorithms are proprietary (they are, in fact, jealously guarded trade secrets that are the constant targets of industrial espionage). Nonetheless, it is arguable that so much of *the news* is akin to the algorithm example. The inter-item and intra-item search functions (Web Content Mining and Web User Mining) are pivotal here, with their emphases on content over structure. Both require a subjective component (human editors) to get things done. Jeff Birkland, product manager for *YahooNews* suggests that "News is far too human an endeavor to rely 100 percent on automation" (Linden, 2004). Professional allegiances notwithstanding, there is something to Birkland's assessment.

Like *news*, and echoing concerns over source and channel, the term "Authority" should be defined in a context-specific and search-specific way. We can return to my car hobby for a moment to understand why. When I need to determine the best way to repair that idle control valve I can reliably locate the necessary information and even quickly compare it with similar procedures detailed on numerous *VW* enthusiast websites, listserves, and blogs. But I would never start at the first level of the Web to do this. Instead, I would seek out a known "base set" (my own or that of someone I trust), access a smaller number of "intelligence communities," and finally settle on a site that has established, relatively speaking, the general ethos of its members and, by extension, that of the site they comprise. The hopeful logic here of course is that the individuals who make up the site are also interested in maintaining a generally positive social climate and, thus, the technical utility of the community. A tacit social contract exists on these sites in the same way one underlies a great sports team or any personal relationship worth the work. The credo is simple: *we are in this for each other* (with, perhaps, a sub-clause: *because we are also vested in our own interests*).

Like any social contract worth its salt, we work to maintain the whole because we are aware of the personal benefits that ultimately trickle down to individual participants: *strength and safety in numbers.* It's a time-honored mantra, and for good reason—it works. In my experience, this sort of normative process eventually forms the foundation of sites like *Corrado Club of America* or *VW Vortex* because the terms are continually cashed out in the interaction that takes place there. A communitarian ethic emerges out of the asking of questions and the provision of answers. After nearly a decade of participation on two car hobby nodes, I have found that the nature of questions, the advice, and the monetary transactions bear this out. Indeed, while there is also a very robust commercial aspect operating on most of these sites, there is also a sense of camaraderie and concern for the broader collective as a virtual community and space that offers relevant, practical, and accurate information and news on related topics. This may be what *Yahoo* hopes to achieve with its distinctive human-machine mix and, more formally, the *My Web* utilities being offered. *Yahoo* seems to be eyeing a similar homeostasis through these symbiotic relations maintained between the system and its users-qua-content.

But beyond this relatively quick narrowing of the search, there is also something special about the whole car repair enterprise that may be impossible to replicate at the scale of a web-based news service. Auto mechanics is a decidedly tactile and concrete pursuit that finds legitimization in the *unmediated*[16] details of my first-person, causal, lived experience. I can locate the best and most reliable authorities in fairly short order because I can jot down or print out a copy of the procedure, walk over to my car, and run through the steps to remove and replace the component and determine if it in fact works as well as my favorite knowledge sources claim.[17]

Then again, when I instruct my freshmen speech students to locate high-quality information on the present status of the debate over a proposed wind farm in Nantucket Sound, the task and the eventual outcome have some affinities to my automotive arts query. For one, the wind farm proposal also has a current and sufficiently concrete ontological status. That status can be determined, more or less, by tapping a relatively localized discussion on the topic that includes narratives and testimony that corroborate (or contradict, if hearsay or urban myth) the findings of environmental impact studies conducted around approved projects of similar type and scale elsewhere in the world. In these instances, local activists, lawmakers, private utility representatives, and energy conservation specialists all contribute to the arguments and discussion. And this all creates a very robust data set. To be sure, all kinds of politicization and politically-motivated obfuscation seeps into the discussion, but reliable environmental and economic impact data related to the proposed generator stations tends to remain constant and readily available.

Now, in contrast, when it came to the matter of several recent Supreme Court nominations things really get interesting. When the primary data for this analysis were being gathered, it was Samuel Alito's nomination or, specifically, Alito's suitability for the job that was up for grabs. Unfortunately, the task of finding reliable data was not nearly as straightforward as the car mechanics and alternative energy queries. A Supreme Court nomination is precisely the sort of news item that lends itself to highly charged, politically biased information on blogs, newsgroups, and even mainstream news outlets that nonetheless get pointed to by automated search engines. So if the medium or technological artifact is the message as McLuhan opined, the content *can* make a difference when assessing the practical benefits and shortcomings of particular technologies. Even an in-depth excursion into the Web regarding Alito in the Fall of 2005 revealed that locating accurate and comprehensible information on his nomination was less than a straightforward enterprise—much less.

Is this a fundamental difference in the nature of news and information about *things* versus *people*? At one level the answer is *yes*, since questions about a thing (or a *process* like a car repair procedure) tend to revolve more around the success or failure of the thing in fulfilling a very particular, concrete, and localized function. If this observation rails against many of the assumptions media determinist hold regarding the causal roles played by technological artifacts, there is a practical sense in which we can determine whether or not something is performing in the manner intended. Questions about people, on the other hand, almost always run beyond their immediate "functions" in both time and space. The seemingly endless postulations surrounding Alito—or, more recently, Sotomayor, and Kagan—are perfect cases in point. News commentators, pundits and bloggers all made their cases as to whether these individuals would be a good judge (or not); someone who leaves their politics at the chamber door; someone who would not (or would) legislate from the bench, and so on. In other words, ethos questions always get bound up in assessments of people, which naturally run beyond the immediate, concrete situation on to more distal concerns about personality and predilection. Establishing authority in each case (regarding information about the person or the thing) is a different enterprise. Both endeavors are further challenged by the immense scale of information now available online.

Recent public debate about the economic impact of illegal immigrants or corporate malfeasance are also good examples that illustrate the authority problem that has emerged online. The immigration question probably has a determinate answer, but it becomes inevitably and endlessly occluded in the speculation and slant, and search engine gaming of cherry-picked data in service to established political and ideological positions. That is to say, the economic, sociometric, even ethnographic data often exists, however, the facts that help determine, say, the economic role of illegal immigrants get buried so far down in a search result list that it becomes a trial, in terms of both time and effort, to locate it. Instead, the *good* data is effectively replaced by the opinion and innuendo (often packed with key search terms) that tends to come first, makes for a "good fight," and effectively maintains news audiences and portal throughput. From both techno-determinist and business vantages, this is precisely what *Google* and *Yahoo* hope for. It is the core logic of the *network effect*—the more inquiries, the better. This also suggests that many problems endemic to political news could be mitigated in part by the more bottom-up, citizen-journalist news communities now emerging.

COMPETING LOGICS: FORMAL STRUCTURE VS. SEMANTIC CONTENT

With a basic understanding of hubs, authorities, and two methods of mining data from the Web now in place, we should consider the way *Google* talks about itself to the world. As part of a technical discussion of the subtleties of the *PageRank* algorithm written for public consumption, the following analysis is offered:

> PageRank relies on the uniquely democratic nature of the Web by using its vast link structure as an indicator of an individual page's value. In essence, Google interprets a link from page A to page B as a vote, by page A, for page B. But, Google looks at more than the sheer volume of votes, or links a page receives; it also analyzes the page that casts the vote. Votes cast by pages that are themselves "important" weigh more heavily and help to make other pages "important." (http://www.google.com/technology/).

Google suggests that *PageRank* relies on the "democratic nature of the Web" and "also analyzes the page that casts the vote." Depending on which sources one consults, the first clause is perhaps perfunctory hyperbole that can be quickly dismissed, or a fact of the digital era that must be taken seriously. Taking it seriously requires a separate socio-political analysis altogether, so the democratic potential inherent to the Internet and WWW will not be addressed here. The second clause is true . . . and a bit misleading. *PageRank* does perform an analysis of separate pages, but the level and kind of analysis is not mentioned. In fact, *PageRank* analyzes the pages that cast the vote (the hubs) but does so by considering the contents by way of an exhaustive keyword-density search.[18] *Google's* system does not perform an intra-page content analysis as a principal function. Based upon the data collected for this study and analyses conducted by the third-party investigators cited here who have performed experimental manipulations of the algorithm, it does not appear to be the case that *Google's PageRank* looks at the relationships between semantic content within discrete documents and pages on the Web as a primary determinant of their relevance, authority, or importance.

With an overriding emphasis on ranking information gleaned from a far-reaching Web structure analysis, *Google* therefore does not consider the semantic content of the page as a central means of weighing its relevance to the searcher's particular interests. This is how we can critique the notion that the Internet is designed to provide us with the information that we *desire* (to borrow Chaffee's original terminology regarding mass media). In much the same way *Amazon.com* pushes book titles in my direction on the basis of the purchasing patterns of other customers who have bought the same book, *PageRank* sends me the Web pages of the minute based upon its latest calculation of the most popular pages related to the search term or phrase I inputted.

While the analysis at this stage might appear to be an effusive way of making the point that popularity does not equal quality, it is necessary to elaborate on these facts to understand one variant of a perennial concern in Artificial Intelligence studies: the *frame problem* (McCarthy and Hayes,

1969; Sperber and Wilson, 1996). The frame problem emerged during early AI research and was later taken up by cognitive theorists and computer scientists. The essence of the frame problem is finding methods of limiting beliefs or *belief states* in machines that have to be updated in response to new information. This suggests that end users of *Google's* and *Yahoo's* news services necessarily involve themselves in a new scale of satisficing from which a precarious kind of bounded rationality,[19] at both macro and micro levels, begins to take shape.

The frame problem, bounded rationality, and satisficing all help explain the significance and nature of the "topic drift" phenomenon. To reiterate, topic drift is part of the reason why a cursory search for news about a new "cat diesel fuel" on *Google* and most other search engines can *drift* and sporadically point to information about fuel produced from the dead feline mammal, as opposed to information on the kind of fuel used in the heavy duty earth-moving machines one was actually interested in. This occurs with *Google* because that system uses an aggregating mechanism that samples the inter-document searches of millions of other people around the world incorporating the same word. There is indeed a certain meaninglessness involved in any statistical aggregating system like this. The machine is, in the philosophical sense, acting non-intentionally. It doesn't know what it believes, or what it is doing. This is not to say that a Web browser with a bit more human input—simply by virtue of there being *some* human input—is necessarily more aware of its actions. However, questions of framing, relevance, and intention, coupled with the nature of news gathering combine to create a new dilemma: *information overload.* In a very accessible article I found at *searchenginewatch.com*, a site devoted to the latest search engine research and development, the phrase used to describe this was "our emerging abundance problem." This is, certainly, another core feature of Digination.

INFORMATION OVERLOAD AND THE
ONLINE NEWS CONSUMER

The idea of too much information does not apply in the realm of machine intelligence in the same way it does when considering human thought processes. While one can eventually reach a point of diminishing returns by increasing the size of a statistical sample, too much information is rarely a problem for modern computational devices. Generally speaking, where sampling is the task, *larger* almost always means *better* for computers. Automated statistical processes—especially when the product of massively parallel digital computations—do not suffer any appreciable performance losses along the way. But we are interested in the practical use of search engines by human beings for the purpose of acquiring news and information about their worlds. In these cases, it is a relatively simple feat to overload an individual person with too much information. This is especially so when a meta search engine like *Google* typically returns between 20 and 30 million results for a query on "Oil Spill" or even 1 to 2 million results for a query on "Oil Spill Preparedness." What happens next depends largely upon the temperament of the person and the nature of the search. If sophistication (in terms of technical literacy), and patience are important measures, then it must be granted that search habits in the public vary widely—and very widely at that.

If there is a trend to be cited at this time, it is that a relatively small percentage of people click beyond the first few pages of results returned by *Google's* or *Yahoo's* news services, or any other search engine for that matter (Resnick and Lergier, 2002). Even when someone is using more narrowed searches with the advanced features now available, most Web searchers still tend to "surf" and "skim."[20] As suggested by the many innovative ways to collect empirical data on Internet activity today, the popular sentiment seems to be that multiple tasks are in-process simultaneously; people tend to be easily distracted; they feel very busy; and subsequently folks are mostly in a rush. Despite the fact that we can get down to manageable numbers, with 100 or 200 results listed, a certain feeling of impatience has diffused into the population along with broadening Internet access.[21]

What I want to suggest here is that one typical behavior of contemporary Web users is to truncate searches prematurely. In the case of *Google*, the search result list forms a hierarchy compiled according to a keyword-density sweep performed by the algorithm. Once again, the keyword sweep is the largest component of *Google's* intra-page content analysis. The primary arbiter in the compilation of search results for *Google* is the larger Web structure analysis that is based upon the search terms themselves. "Google is certainly mindful of the importance of community and user-generated content (e.g., APIs, Blog Search, Orkut, etc.), but the company appears focused on developing tools and applications to aggregate that content from third-party sources rather than create an environment in which its users create the content themselves. Future research will have to interrogate these dynamics much more rigorously, but one immediate point of divergence between *Google* and *Yahoo* seems to revolve around this "social" or "community" issue.

Web Structure Mining or *Web Structure Analysis* are two terms used today as a way to describe and enable "social networking" online. However, for reasons described above, this is something of a misnomer. The more appropriate vehicle for social networking would be a combination of WCM (the intra-document analysis method already discussed), and *Web User Mining*, an approach that was being paid much more attention to by *Yahoo* than other mainstream Web search engines at the time of this writing.

Web User Mining (WUM) is a third information search and assessment approach that centers upon the analysis and assemblage of a search profile for each user. Essentially, this amounts to an encoded facsimile of the user's online behavior based upon search and storage activity. Among mainstream search engines, *Yahoo* was still spearheading the user-mining approach in 2007. In WUM analysis, a user's bookmark, history, and cookie files are combed in the construction of an emerging database. Various desktop tools are in beta testing that also mine the user's entire hard drive contents to assist Web search activities.[22] A unique mix of objective and subjective processes is at work in such cases. WUM is also machine-driven, but it is designed to base its search offerings upon the user's personal stock of information. This is essentially how *point of purchase* technology works at the grocery store. When I place my discount card on the conveyor belt and the cashier scans that quart of milk, a bunch of grapes, six bananas and a pack of four chicken breasts, the system is keeping track of my evolving purchasing pattern and building my personal "consumer profile." Based upon my selections that day and all days prior at any outlet connected to the same system, the register spits out my receipt along with a number of coupons and special deals that it guesses I might be interested in. The grocery store system mines my consumer profile, establishes a very small root set of items on offer throughout the entire store and predicts (and even prompts) my subsequent purchasing behavior by offering a wider, yet still constrained base set for my consideration. This is very different from the WSM strategy that forms the core of *Google's* search logic.

In addition to a structural analysis of linkages between discrete pieces of Web content, *Yahoo's* browser system blends content mining and user mining to obtain a smaller base set to work from in an effort to provide the user with tighter (i.e., more relevant, authoritative, and legitimately tailored) results. If I am still allowed to conceive of news gathering today as an activity that can provide the constituents of a population with information to help them make informed assessments about, and choices in, the society of which they are a part, we can say that *Google's* universe is certainly larger and more differentiated. This may be what the *Google* spokesperson is alluding to above when he says that the Web has a "democratic nature." But if participating in a democracy is about exercising one's ability to make free and informed choices from a broad set of options, then what *Google* offers up from the Web may have already surpassed that crucial point of diminishing return. Other observers suggest that the numbers borderline on insignificance in this particular matter, however a slightly larger proportion of *Google's* consuming public (relative to *Yahoo's*) may have no interest in the news items being offered up, or even an awareness of that item's relevance to their own lives. On the other hand, it could be said that *Yahoo's* somewhat more subjective base sets and, subsequently, its smaller data universes lead to a kind of artificial narrowing that in turn prompts (or at least enables) a cognitive narrowing in its user base. In terms of substantive differences, however, I have to admit that these remain open questions in need of much more deliberate investigation.

According to Fallows (2005), two features of *Yahoo's* approach set it apart from the way *Google* has been doing things. The most recent version of the *Kleinberg* algorithm underlying *Yahoo*, with its primary emphasis on content mining (and near equal consideration of *Hubs* and *Authorities*) is the first. The increased attention *Yahoo* is paying to personal user profiles via user mining and the maintenance of a social network is the second. Together, these features may explain the somewhat tighter, and sometimes more relevant content higher up in Yahoo search results. Bona fide social networking was becoming a core institutional value at *Yahoo* in 2005, exemplified in their *"save to my web"* option.[23] In the face of *Google's* larger and solely machine-based aggregations, *Yahoo* is betting on the human factor to ensure legitimacy and relevance for users of its search systems and (perhaps most importantly for stockholders) allegiance to the *Yahoo* brand itself. As *Yahoo's* chief technology officer, Farzad Nazem stated in 2005: "We're really about getting the average consumer to move their lives online" (quoted in Fallows, 2005). Not to be outdone, *Google* has a similar plan in mind:

> Google seeks to become the gatekeeper for not only the public Web but also the "dark" or hidden Web of private databases, dynamically generated pages, controlled-access sites, and Web servers within organizations (estimated to be tens or even hundreds of times larger than the public Web); the data on personal computer hard drives; and the data on consumer devices ranging from PDAs to cell phones to iPods to digital cameras to TiVo players. Google's founders understand the scale of the opportunity. Larry Page recently said, "Only a fraction of the world's information is indexed on our computers. We are continually working on new ways to index more . . . Thirty percent [of our engineers] are devoted to emerging businesses." And Sergey Brin once told Technology Review's editor in chief, "The perfect search engine would be like the mind of God" (Ferguson, 2005).

CONCLUSION

Google, AOL, Yahoo, YouTube, and *Wikipedia* are all routinely referenced on Jay Leno, David Letterman, Jon Stewart and Jimmy Fallon to name a few popular media conduits. *Google* has become one of two, and increasingly the sole Web portal named on CNN, Fox News, and other mainstream news outlets. The powerful moniker has migrated elsewhere too. Google is referred to as the default search engine in clothing, food, furniture, real estate, travel, beer, sports, and automobile advertisements on TV, in newspapers, and on the radio. Of course, from the perspective of those "in the know" the appearance of search-engine reduction or condensation (short of monopoly) is recognized as stratagem, and really is just good business practice. Indeed, it becomes a predictable expansion of corporate market valuation that has little to do with the enhancement of news production and consumption. But for anyone who confronts the technology with a less critical eye, the common conflation of *Internet* and *Web* is harmless compared to the popular trend to equate the *News* with *CNN*, or the *Internet* with *Google*. This condensation of search technologies should be an urgent concern for communication theorists and practitioners. But in addition to those thinking and writing about mass media effects, the contemporary status of political action, citizenship, news consumption, knowledge-building and related epistemological issues, such condensation also raises a host of pressing questions about emerging processes of persuasion and influence.

Katz and Lazarsfeld's seminal book *Personal Influence: the part played by people in the flow of mass communications* (1955) made the 2-step flow model of influence a steady fixture in mass media studies throughout the second half of the twentieth century. While it remains a core text in the field, the ideas in *Personal Influence* have been updated and extended in numerous books and collections (cf. Simonson, 2006). Above all perhaps, Katz and Lazarsfeld posited that mass media content is filtered by opinion leaders, who then pass on what they deem most relevant to the people around them.

The popular explosion of search-engine technology in the mid-1990s has substantially blurred the distinction between all three nodes of the influence process noted by these authors: (1) content; (2) opinion leaders/experts/authorities; and (3) the public. If certain corners of the commercial realm have capitalized upon the confusion between *expert* and *entrepreneur* we are seeing online today, it is an understatement to say that academic and popular opinion are now also experiencing a sort of upheaval in this regard. This is especially so where *news* is the information genre under consideration. In 2006, more than 68 percent of American households maintained either dial-up or broadband connections to the Internet. In 2007, Internet use among Americans was as high as 79 percent, with 47 percent of Americans enjoying broadband access (*Parks Associates, thealphamaker.com,* and *pewinternet. org*). So as the cost of a *netbook* moves steadily down toward $150 (or less), and as Wi-Fi signals continue to open up around the country, the numbers of

Americans connected to and experiencing the multifaceted effects of Digination continues to rise.

And so with all of this the definitions of *authority* and *expert* are up for grabs as the parsing mechanisms at the core of *Google* and other leading search engines direct the ebb and flow of the emerging data sphere. If human knowledge, like human *being*, is always in the process of becoming, we enter an unprecedented situation today where non-human agents are defining and delineating our collective knowledge, and our wider information ecologies. Along the way, popular impressions of significance and relevance, notions of authority, and even a basic understanding of what it means to say something is "true" move further beyond human influence.

Do immigrant workers consume more resources than they produce, or will they eventually make a positive contribution through mechanisms like social security and other taxable inputs into the capital stream? These are important questions that actually do have a relatively clear set of answers. However, even though there is an embarrassing abundance of information on this topic online today, the question of immigrants remains a hotly contested and underdetermined issue in a way that might allow viewers of a debate about immigration policy on *Meet the Press* to walk away with a relatively transparent understanding of things.

What we find is that certain structural features of the two most popular search engines supplying us with information today lead to different perceptions and understandings of source, channel, quality, authority, and, by extension, truth itself. Given the dizzying manner in which these aggregating systems produce result lists, questions as to what constitutes a reliable source, or an authority in any given topic area, proliferate. Hopefully, I have demonstrated to the satisfaction of the reader how, in trying to establish the "facts" surrounding politically charged issues like immigration, or a Supreme Court nominee's bid, or the status of the war in Iraq, the hope of finding an *authority* also seems a bit naïve. Indeed, the whole notion of an authority seems to reduce to little more than a distant, fading, platonic ideal.

If mainstream news outlets are now enabled by the Internet to offer us *all the news all the time*, they remain bound by the commercial imperatives of any industry seeking to turn a profit. Regarding the ongoing conflict in Iraq for instance, this corporate function of the mass media has been steadily critiqued over the last decade and a half (Aday, 2005; Hatchen and Hatchen, 1992; McKenna, 1999). No doubt, the current confusion online is abetted by other variables too. Corporate interest, military censorship, and even self-censorship among journalists has rendered news reporting a hit and miss endeavor for the news consumer seeking information on- or off-line today. It is the online components of news organizations that are becoming their central nodes of production and dissemination, and this is occurring at a time when more and more automated systems intervene in news production and consumption processes.

Emblematic of these changes, *Google* added "Hot Trends," an automated Web query-tracking feature in May, 2007 to replace "Google Zeitgeist," its manually compiled list of the most popular search terms that scanned all terms, including news related items. As the public's expectation of quick and ready access to news and information accelerates in step with the latest search technology trends, traditional news criteria like relevance, accuracy, and depth drown in their wake. The wider effects of these trends on the day-to-day doing of journalism, the technical production of news, and the general practices of news consumption and processing by the public have yet to be determined. They will need to be investigated further.

The Electronic Commonwealth: the impact of new media technologies on democratic politics (1988) posed a number of intriguing questions that centered on the way top-down portrayals of political actors in the mass media get recapitulated by smaller news outlets. These standard procedures can skew the public's understanding of those individuals and the issues and events they are associated with. There is a long tradition of media effect research that considers the way human designers in positions of power have, intentionally or not, directed the public's knowledge and awareness of events in the news with formal analyses going at least as far back as Walter Lippman's *Liberty and the News* (1920). But if the future of news looks bleak to so many for a variety of reasons already discussed, there are some glimmerings of hope on the horizon. There are effective "work arounds" out there.

In the face of *Google, Yahoo,* and other leading online news filters, one idea is that "on-the-ground" citizen journalists embody part of a solution to the puzzlement surrounding authoritative news and information today. If they can obtain a basic proficiency in Web search gaming and can then attract and hold audiences by virtue of their superior content and, perhaps above all, good writing and research, Blogs, Vlogs, *Myspace,* and *YouTube* all represent test beds where even the lone muckraker can help tip the scale away from conglomerate-centric news search results. It even seems reasonable to suggest at this stage in the morphing of our *news mediascape* that more opinionated, even explicitly partisan news sites would be an enhancement over the bland policy allegiance that typifies the surprisingly narrow editorial variances and "trickle-down" dissemination of mainstream news today.

Almost twenty years after *The Electronic Commonwealth,* and even as we begin to understand the basic structure underlying a more bottom-up model of information assessment and retrieval, this new era of infobots and AIs roaming the Internet and World Wide Web has changed everything. One thing seems clear: the parsing technologies incorporated in part or in whole into our most popular search engines, as well as concerns over the skewing, blending, and blurring of *sources, channels, authorities, experts,* and *news* will only increase in depth and scale. As we look further into our digital domains for new ways to gather information, to have knowledge, to be aware, to *really know about something*, we may inadvertently discover a little bit more about what it means to be human today, for we are still very much *under construction.*

NOTES

1. Statistics vary according to measurement technique and source, but most analysts put Google ahead of AOL, MSN, and Yahoo by factors ranging from 3 to 5 orders of magnitude (Neilsen Netratings, and 1Cog.com) for the US market. Global Google usage accounted for 56.9 percent of all searches conducted in September, 2005 according to onestat.com.

2. Berners-Lee is, by most accounts, the man who invented the Web (if that moniker could be attached to any one person). I think Berners-Lee remains hopeful that "hidden amid all [this] data is the key to knowledge about how to cure diseases, make more money, and govern our world more effectively" (Berners-Lee, 2007). However, he also asserts that "the technical tools and social practices that shape the way we manage, share, integrate and analyze this under-used treasure trove are sorely out of date" (ibid). While there may be some problems with his neutral theory of the Web, Berners-Lee continues, with work on his "semantic web" project, to be most interested in creating better connections between human minds, not machine-driven, functional approximations of human minds.

3. There is talk of an eventual standard because the Internet is a domain largely managed by and organized around interests that are primarily commercial in nature. Despite much lip service about decentralized content and authorship online, advertising support and the general "ad subsidy" concept continues to hold sway on the Web in similar proportion to that sustaining commercial print media, radio, and the still massive television industry. This is one reason why utopian visions of the networked society proffered by Negroponte, Benkler, and more recently Sunstein always fall short of the reality we experience. If some of us may be successful in avoiding the trappings of banner advertisements, spam, splogs (i.e., spam blogs) and the like, the continued success of Web-based marketing suggests that there has to be a large enough proportion of the population who does not manage to avoid or ignore these features. As opposed to the mapping of demographics with more conventional media, the difference in strategy online centers upon the fact that in order for advertisers to be able to reliably identify, track, and target potential customers, they must at least possess a general understanding of the referencing, indexing, and forwarding mechanisms of the technology. When a company builds a website (depending on how they order and list keywords and terms in the site's underlying HTML code), they may discover that their site is systematically overlooked by potential customers

using the news services run by Google or Yahoo, or some other search engine. A side effect of the commercial dictate to raise "click visibility" of commercial businesses, organizations, and news sources is the emerging industry of search engine gaming. Content farms are a more recent instantiation of the Internet's particular media logic. This is also, in part, why many partnerships are being formed between Internet service providers, traditional news and media outlets, and leading browser and portal brands like the recent corporate linkages made between Microsoft and NBC, the Washington Post and Yahoo, and AOL and Google.

4. On this view, information access should be conceived of something very much akin to a civil right. Promises, and some efforts, to close the "knowledge gap" or "digital divide" between the information haves and have-nots by many of the above mentioned notables suggest that we are working to solve this new problem of the digital age. Indeed, if a representative system is based upon the voice and will of the people, and if the people should be well informed in order to participate in an informed and deliberative way in the political process, then the provision of information required to fulfill that civic duty should be universal. Google's Brin and Page, for instance, have taken up a significant portion of that charge in their promise to digitize the world's print collections and provide access to all with a nominal Internet connection.

5. While the objectivist, referential worldview embedded in so many optimist visions is suspect in and of itself, the idea that confusion will somehow–and quite naturally–give way to clarity if enough data is made available seems suspect, or at least naïve. However, this image of an emergent moral and ethical component coming forth persists. Negroponte's, Mitchell's, and Benkler's stories all allude to some idealized conception of verifiable fact, even truth guiding content providers, and that Web users (including news consumers) can then occupy an objective vantage that somehow cancels out their unique and highly skewed subject positions. This is not to mention the intentional obfuscation perpetrated on listserves, blogs and websites in the service of vested or special interests. All of this, furthermore, casts doubt on the emancipatory aims associated with Web2.0, and seems to be an even less tenable foundation upon which to base a "digital democracy."

6. Manuel Castells' Vol.I: *The Rise of the Network Society* (1996), Vol.II: *The Power of Identity* (1997), and Vol.III: *End of Millennium*.

7. A number of treatises describing the strengths and shortcomings of the "many minds" approach to knowledge creation have been published in the last several years. Cass Sunstein's Republic.com (2002) centers upon a concern that the Internet and, more generally, the network effects bound up in it contribute to the progressive disintegration of the public sphere and diminishes our hopes of realizing a digital democracy. However, Sunstein's *Infotopia* (2006), and Yochai Benkler's *The Wealth of Networks* (2006) bring the good mood back as both engage in broad analyses of the positive potential open-source software and applications, and Web2-like platforms hold for our collective social, political, and economic futures. Sunstein and Benkler converge on many of the same implications and hopes connected to the creative commons and other networked knowledge structures emerging today. These

include comparative discussions on the benefits and shortcomings of searching by pure algorithm vs. searching with remote human judgments intervening along the way (the essential distinction investigated in the present analysis). Both Sunstein and Benkler also frame the problems and questions of search approaches and theories in broad terms that take into account the strategies and costs of information searching, and the implications for social, political, and economic life.

8. Jon Kleinberg gleaned considerable insight into the structure of the Web itself while doing research on RNA as a graduate student. "Kleinberg eventually noticed that high-quality sources, such as repositories of research papers and home pages of prominent researchers, tended to be referenced by high-quality reference pages, like course home pages and carefully compiled bibliographies. Thus, Kleinberg realized, useful pages on a topic are of two types: Authoritative sources contain information on the topic; reference pages, which he calls 'hubs,' provide links to sources. To Kleinberg, this suggested that the links in the reference pages encode judgments about the quality of the source pages; the quality of the reference pages, in turn, can be inferred from the quality of the source pages they link to" (Robinson, 2004; p. 2).

9. IMAs automatically discover related material on behalf of the user by serving as an intermediary between the user and information retrieval systems . . . [b]ecause IMAs are aware of the user's task, they can augment their explicit query with terms representative of the context of this task. In this way, IMAs provide a framework for bringing implicit task context to bear on servicing explicit information requests, significantly reducing ambiguity. IMAs embody a just-in-time information infrastructure in which information is brought to users as they need it, without requiring explicit requests (Budzik and Hammond, 1999; p 1).

10. "More time" typically means several hundred more milliseconds when massive parallel processing such as this is at work.

11. "At the moment [Google's] search engine offers up tens of thousands of 'hits' in response to simple entries such as 'Iraq,' which lead to news websites. These are ranked either in order of relevance or by date, so that the most recent or most focused appear at the top of the huge list. This means that articles carrying more authority, say from CNN or the BBC, can be ousted from the first page of results, simply because they are not as recent or as relevant to the keyword entered in the search line" (Fox, 2005). Inspired perhaps by Berners-Lee's original vision, Google Scholar is one attempt to restore a research-oriented base set (and basis) to the Web. Otherwise, increasingly, one has to sift through larger and larger numbers of pages to locate relevant information.

12. PageRank is a commonly used algorithm in Web Structure Mining. It measures the importance of the pages by analyzing the links. PageRank ranks pages based on the Web structure. Google first retrieves a list of relevant pages to a given query based on factors such as title tags and keywords. Then it uses PageRank to adjust the results so that more "important" pages are provided at the top of the page list (Xing and Ghorbani, 2004; p.2. Italics added).

13. Amazon's automated references stand apart from their reader reviews. The written reviews are a closer approximation to Web User Mining (WUM), the parsing method discussed in this section.

14. The occasion for search result "gaming" and other underhanded forms of result list tinkering is ever-present as hackers, computer engineers, and entrepreneurial spirits in general learn the hidden subtleties of the various algorithms roaming the Web today. But a more legitimate kind of gaming also manifests whenever business relationships are established online. Indeed, standard business practice also prompts Google, AOL, and Yahoo to steer viewers to its affiliated sources and channels by packing pages with key terms and phrases in order to fulfill advertising contracts with clients.

15. When we go online and engage wholly automated systems like Google or Metacrawler, or MS Explorer in a Web search, human intentional activity is periodically suspended and interlaced with non-human activity, as the browser proactively retrieves files from a distributed collection of information repositories.

16. It can be argued that all of our interactions with the world are always and already mediated by the particularities of our senses as well as the subsequent idiosyncrasies of our individual, phenomenal experience (for instance, I'm mildly colorblind and have a tin right ear). But the present discussion hinges on the formative characteristics of communication media themselves, so we can say that such an event is not being mediated by any technical apparatus other than the wrenches, sockets, and pliers necessary to do the job.

17. As with political news blogs, user-groups like these also run the risk of descending into "groupthink." However, a rabid (if largely innocuous) kind of "brand allegiance" is typically the worst that can come of it on the average car hobby site. Granovetter's *Strength of Weak Ties* (1973) laid the foundation for an argument that suggests how in mediated interactive contexts we can acquire high quality information from people we do not know.

18. Unlike many news aggregators that simply "scrape" headlines and links from news sites, Google's news crawler indexes the full text of articles. (Sherman, 2002).

19. Bounded rationality is meant to describe the kind of "thinking" that occurs when the world is perceived to be too complex to understand. In these cases the agent behaves so as to maximize its goals in the present context or as far as its available resources will allow. "Here, bounded refers to inherent limits on rational thought, depending on the organism and its environment. Decision strategies that hinge on mere bits of well-chosen information about one's surroundings pack a surprisingly powerful wallop, especially when time and knowledge are in short supply, according to the bounded-rationality view" (Bower, 1999).

20. A host of "click counters" and search-tracking companies are now operating for free or for profit that support this observation. An intriguing computer art installation called the Listening Post that was touring select galleries over the last several years that records and audio-visually represents "the 'collective buzz' of the world mind" (Gant, 2006). An earlier version of the piece I was fortunate enough to experience back in 2000 was a more simple audio installation set up to represent (through the

registering of tonal and pitch changes) how many people were visiting a website, and how deeply (in terms of "clicks" below the surface), at any given moment.

21. A common reply to this observation is that human impatience is nothing new. If we aren't born with an appetite for the next fastest thing, we are seeing a new kind of intolerance for delay in the Digital Age. Awareness of the very systems being described in this study is part of the new phenomenological experience we are having in relation to our machines and the data-scape they offer us. The notion of a virtually bottomless news and information repository carries along with it a certain theory of what is possible, but the real world has a tendency to muddle the translation from theory to practice. Indeed, there is a kind of conceptual disconnect between the way we might often think about the news that is available to us online and the way we actually go about obtaining it. It is very likely that Yahoo and GoogleNews can, in most instances, probably deliver what we desire and ultimately need in order to perform our roles as citizens. Part of the problem is that these systems are often unable to do so in the face of individuals' real or perceived time constraints.

22. "Until now, competition in the search industry has been limited to the Web and has been conducted algorithm by algorithm, feature by feature, and site by site. This competition has resulted in a Google and Yahoo duopoly. If nothing else were to change, the growth of Microsoft's search business would only create a broader oligopoly, similar, perhaps, to those in other media markets. But the search industry will soon serve more than just a Web-based consumer market. It will also include an industrial market for enterprise software products and services, a mass market for personal productivity and communications software, and software and services for a sea of new consumer devices. Search tools will comb through not only Microsoft Office and PDF documents, but also e-mail, instant messages, music, and images; with the spread of voice recognition, Internet telephony, and broadband, it will also be possible to index and search telephone conversations, voice mail, and video files" (Ferguson, 2005).

23. Yahoo began beta-testing its Save to My Web tool with early users of Yahoo's MyWeb2.0. This option lets users add Web pages to their stored pages on MyWeb with a single click.

Chapter 7

The Sound-Tracked Lifeworld

"All art aspires to the condition of music"

—Walter Pater

Even as the high definition image gathers so much attention on the cultural front, with big, flat plasma and LCD screens flashing HD and Blue-Ray content, the personal experience of music and sound remains one of the most whole and complete human sensory phenomena. Music fills in the gaps of experience. It oozes into the corners of consciousness and can settle us into a feeling, and a lifeworld in a way that makes even the richest visual experience often pale in comparison. The term lifeworld has been employed in fits and starts throughout the book so far but in this chapter and the one that follows it plays a central role in understanding everyday media use. So it's high time I describe the philosophical concept in a bit more detail.

The *lifeworld* (from the German *Lebenswelt*) is generally used to describe an environment, domain, or totality that is perceived to be, a priori, self-evident or given to the subjects doing the experiencing, and should be the starting point of any sociological, philosophical or epistemological inquiry. The lifeworld, in Edmund Husserl's original (1936, 1970) formulation, is a collective experience, but I will bend this piece of the definition slightly to demonstrate how some of the most significant, meaningful, and behaviorally consequential moments of daily life are increasingly being experienced in isolation (even if collectively created). Indeed, a kind of systematized solipsism is one of the most powerful side effects of the media described here.

In filling out her answers to a survey regarding the way one might explain the personal meaning of her *iPod* to a proverbial Martian, Heather, a thirty-something urban professional living in Boston, had this to say:

> My *iPod* gives me a way to enjoy my favorite music, whenever and wherever I want. I don't particularly care for "mainstream" music these days and so a lot of the artists I listen to are either completely independent or are very, very obscure . . . I wouldn't be able to ever hear them on the radio. The iPod is small, it's light-weight, and since I'm usually caring a lot of stuff with me, those design aspects are important to me. Also, because I am from the Midwest and I grew up always having plenty of personal space, even in public, I use my iPod as a way to maintain distance when I can't physically have it. On the T when I'm crammed in there with a million other people, I can turn it on and they all fade into the background and I can concentrate on my music and on me. I play some Mulvey and it keeps my morning commute, especially, from being a stressful part of my life. On the days that I don't have my iPod, I end up stressed, annoyed, and unhappy for most of the day . . . so I credit it with adding to my quality of life by giving me the "mental" space that I need to get through a day in the city.

What can a sudden track change or random switch from *Peter Mulvey* to *Rage Against the Machine* do to one's sense of self and one's feeling toward the myriad others they come in contact with throughout the day? This chapter is about the phenomenological effects of music and sound and the ephemeral lifeworlds that result from both the conscious and contingent creation of different sonic environments enabled by small portable digital music devices that have become intimate and essential pieces of apparel—and of the self—for so many today.

THE PHENOMENOLOGY OF THE MOBILE LISTENER

Phenomenology was a pragmatic departure from the highly abstract and conceptual-analytic trends in fashion in the late nintheenth century. In keeping with Edmund Husserl's original formulation, contemporary phenomenologists are guided by a philosophical perspective and method of inquiry based on the idea that reality itself—the lifeworld—consists of objects and occurrences as they are perceived or understood in human consciousness. Phenomenological arguments and outlooks lay at the root of much medium-theoretic thinking, especially the Media Ecological tradition that includes McLuhan, Ong, Gumpert and Cathcart, and Meyrowitz. With a combination of hand distributed surveys, some follow-up e-mail correspondence, and a series of small focus groups and in-depth interviews I was able to cull some incredibly detailed phenomenological data from my research participants.

With this data I was then able to draw out a robust picture concerning the meaning and function of portable digital music devices today.

Listening to certain anthems from my college days brings me back to that time and place with a vengeance. For just about any social moment in space and time there is probably a theme song or personal soundtrack that helps conjure it. Music reminds and binds us to these spatial-temporal contingents like few things can. It does seem to be the case, as Walter Pater opined late in the nineteenth century, that all art [still] aspires to the condition of music. Music seems, above all else, to be about feeling. There is something in the emotional response to music that moves us, but it is more than just an emotional response. It is pure aesthetic—or as close to this as any single media experience can get.

In this chapter I explore some of the various ways personal digital music devices (PDMDs) enable the reconfiguration of perceptual space within individualized experiential contexts. These devices afford their users a relatively high-resolution sound space that, when tuned properly, often *re-places* the perceiver both spatially and temporally. In so doing they have the capacity to fundamentally recast the nature of personal experience. By far the most popular device, Apple's iPod, has been shown to re-create the listener's acoustic space, as it remediates the surrounding context by filling in the parodic "cone of silence" in Mel Brooks and Buck Henry's *Get Smart.*

A VERY BRIEF HISTORY OF PORTABLE
AUDIO TECHNOLOGY

Going mobile with on-board audio changes everything. While we could begin with "portable" record and reel-to-reel tape players, or the portable transistor radio of the late 50s and early 60s, the history here is, in fact, very brief if we limit ourselves to the sea-change that occurred with the advent of portable personal sound programming. The portable cassette tape player brought this capacity to the consumer for the first time in the late 1970s. The device was invented by engineers at SONY corporation. SONY's *Walkman* line (and all of its copies) radically altered the nature of our soundscapes. In much more nuanced fashion than the car audio system, with a pair of stereo headphones these devices made the private mobile soundtrack a practical possibility for the first time in history. The rapid diffusion of these relatively inexpensive little machines also transformed the look and feel of the daily commute, the jog through the park, not to mention office spaces throughout the world. To be sure, due to the basic multitasking abilities our sense of hearing has always allowed, the Walkman brought a new reality to the notion of *being all alone together* (Moebius and Michel-Annen, 1994).

The portable compact disk player followed in the late 1980s, almost exactly one decade after the first SONY Walkman cassette player became available to the consumer. The CD significantly enhanced sound fidelity for the listener on the move but temporality suspended the programmability function. However, it was not until the MP3 audio file type was invented in 1991 that the possibility for the reliable, convenient personal recording and storage of audio became a reality—and then it was the portability feature that needed time to catch up. The ability to program personal music lists in portable digital formats occurred five or six years later in lesser-known portable digital formats like DAT, DCC, and MD) but these forms and formats always struggled in the consumer market.

I obtained a remanufactured ("B-Stock") portable DAT player from a friend working at SONY's recording media office in NYC in 1993. Unfortunately the cool little device ate through batteries much faster than I could afford. And then, about a year into owning it, the thing only seemed to work when it wanted to, eating tapes or emitting a high pitched squeal (or both) when the humidity spiked even a little.

It wasn't until the SaeHan Information Systems' *MPMan* portable MP 3 player hit the scene in 1998 that things really started to happen. Several of the usual suspects tried their hand at these early portable devices in the late 90s, but only to lukewarm success. Creative Labs comes out with their popular Nomad Jukebox, a pudgy little portable device, at the turn of the Millennium. In October of 2001 Apple introduces its first generation iPod. And then, as they say, the rest is history. While virtually all mobile phone manufacturers now produce models with onboard MP3 players, the iPod still sets the standard for user interface, battery life and memory capacity, though programmability has been stymied for some by Apple's proprietary *iTunes* protocols.

DISCRETE SOUND SPACES

In most urban centers today we still experience something R. Murray Schafer (1973) dubbed "sonic overpopulation." More than 30 years after Schafer coined the phrase we still suffer from an over-abundance and intensity of sound. Sound pressure levels in and around cities have increased over time as two of the greatest noise-makers—jet and internal combustion engines—have proliferated. Today a discordant symphony of sound confronts the pedestrian in an urban environment. "There is so much acoustic information that little of it can emerge with clarity. In the ultimate lo-fi soundscape the signal-to-noise ratio is one-to-one and it is no longer possible to know what, if anything, is to be listened to" (Schafer, 1977, p.71). As Schafer hints there is often good reason to try and cut out the cacophony and envelop oneself in a private sound-space or soundscape.

Most of the ambient sounds we experience on a daily basis can be effectively neutralized by $20 silicone/latex in-the-canal earphones. For a little more than $100 today, active noise-cancelling phones nearly obliterate the ambient acoustic environment. Now we can quite literally *turn up the silence*, though most people I know rarely use their noise-cancelling headphones in this way. I typically use my *Philips* noise cancellers in airports and on airplanes to bring down the *noise floor* at least a few points on the decibel scale. The noise floor is a sonic measurement of the quietest a given environment can get—a sort of acoustic baseline. My noise-cancellers are designed to cut out 75 percent of the ambient sound, theoretically reducing the noise floor by a factor of 4. (They don't actually work that well, however).

Unfortunately, reducing the noise floor is not what automobile sound systems are designed to do. To the contrary, the average system raises the local noise floor by attempting to overcome the ambient sounds through a kind of brute force. These systems are designed to be loud enough to out-sound road noise, wind noise and just about any other noise one can think of. Indeed, you just turn up the volume. Most late model cars come equipped with a fairly impressive audio system installed at the factory that will do the trick. But there is something acoustically and phenomenologically distinct about cruising around to our favorite sounds and ambulating about to them in bipedal fashion. Both certainly have their pleasures. Where volume and a literal tactile experience is concerned, however, the auto-audio experience may win out. The average car stereo system today includes at least one four-channel, 50 watt, line-level amplifier, with six speakers and, more often than not, a small subwoofer under a seat or hanging in the trunk. For about $500 more (with some people easily spending as much as $5000 for a competition-grade system complete with sound dampening material), a high-end sound system with upwards of 300 watts and very low distortion levels can be installed by *Best Buy* or an after-market mobile audio specialist for a nominal fee. And again, with such systems it is very easy to completely occlude the ambient sounds emanating from the world. One simply turns up the volume—all the way to "11" as Christopher Guest suggests in *Spinal Tap*. With a single 8 or 10 inch double voice coil sub-woofer and a low-pass cross-over, sound becomes tactile. At a sound pressure level of 130 decibels, a sound wave oscillating at anything less than 200 hertz rattles the spine.

There's also the ongoing frenzy to produce the most sophisticated surround sound systems for the home theater. It has been nothing short of a technology explosion in both the car and home audio markets over the past twenty years, yet all of this R&D often just ends up approximating the experience of listening in stereo on a $50 device through an average pair of earphones. And so I want to focus on that mobile, pedestrian sound setup that has become so ubiquitous today as it is emblematic of our Digination. I'll suggest that is the combination of using a PDMD as a pedestrian that potentially short-circuits those portions of our civic spaces and places sometimes referred to as the "public sphere." As one of my better-mannered respondents explains:

> I have noise-blocking earphones, so people can't hear what I'm playing and I can't hear what they're saying. But if someone approaches me to speak, I always immediately remove the earbuds and let the conversation take a front seat to the iPod. I don't use random or shuffle that often; I prefer playlists instead. (Jed)

As if a direct affront to Jed's rare etiquette these days, a recent iPod advertising campaign features silhouetted figures flailing about to their iPod and iTunes. These ads are striking for a number of reasons. First of all they're silly. I'm pretty sure I've never seen anyone dancing around like the folks depicted in those colorful ads. But beyond this, the ads speak to something real that is happening in our cities and towns. And it is a haunting sort of trend. While the number of bodies may not be diminishing to any appreciable degree, we are witnessing a hollowing out of the population. Apple's suggestion is that these are individuals caught in the middle of impromptu dance moves—the ads, however, also and ironically point to the disappearance of the person-as-social-agent. And here the whole "social-minimizer" or "free-rider" scenario takes on new meaning.

PDMD-equipped, the social minimizer also becomes the programmed urban automaton. If there is always going to be exceptions to the trend, being plugged-in tends to short-circuit our ability to participate in the living cityscape, the ongoing flux of the social environment. In his theory of *derive* or drifting, Debord (1981) argued, without comment on the social significance of the just-arrived SONY Walkman, that people should resist the tendency to fall into programmed routes and patterns of movement through the city and let the spontaneity of personal whim, of the closed, contingent, and detoured route set off happenstances, including all variety of unpredictable social encounters. I wholeheartedly agree with Debord's suggestion, when there's even a little spare time, and point also to De Certeau's (1984) notion that individuals finding their way through the streets constitute the city analogously to the way individual speech acts constitute language. I would even go so far as to argue that it is on these contingencies of pedestrian movement that the growth of a public depends.[1]

In addition to effectively removing the social actor from the civic sphere (and just about any social context one can think of for that matter), unless one has a good handle on American Sign Language social intercourse is all but nullified. Respondent Jed is a rare bird indeed. However, certain aspects of the self can also be emboldened and bolstered when listening on a PDMD, and a collection of notable experiential attributes seem to be resurrected from our collective past in the process of listening. I'll discuss some of the key features of this sonic sensibility here. But first let's consider a couple of snippets from the popular e-press to help frame the issue:

> A marketing report lands on my desk; its central theme emblazoned across its cover: "*Is this the most 'Me' focused generation ever?*" The question is rhetorical and it is a bold claim to make. Are people really more egocentric than the children of the 1960s and 1970s, Tom Wolfe's *Me Generation*, who leapt away from postwar austerity towards pointless material comforts, indulging their drippy narcissism with Californian singer-songwriters, ersatz analysts and suspect versions of eastern religions? "Don't knock masturbation," said Woody Allen, the era's most acute chronicler: "it's sex with the one you love."
>
> Yet the report makes a confident case. It is based on three main ideas. The first is that, despite being better "connected" than any other generation in history, today's young adults are actually less engaged with the wider world than ever. Even the icons of youth . . . hooded tops, iPods, mobile phones—sever users from the community at large. They are potentially better informed about the world, but care about it less (Aspden, 2005).

While I'll contend, in the next chapter, with this argument that young people today are better informed than the community at large, it is the *"I"-ness* of the iPod generation that warrants further inquiry. There is something seductive about these new sound tools. Indeed "who could resist the invitation of those dainty headphones? They gleam . . . and entwine themselves around heads all by themselves (Kracauer, 1995; p.333).

> Descriptive as it is of the thin white wires that are now so commonly seen snaking up from bodies, bags, and pockets of iPod users, this image was actually constructed by Kracauer in the 1920s in reference to the headphones of a Wireless radio set. Despite having been written decades ago in reference to much older technology, Kracauer's quip is still uncannily appropriate for the iPod and its distinct white headphones.
>
> [C]laims are made about its effects—from predictions that it will spell the death of radio to proclamations that it will irreversibly alter societal relations. However, the iPod fits into a lineage of communications technologies tracing back to the Walkman and the radio before that, and many of the issues now raised with regard to the iPod have been raised long before with regard to these previous technologies (Byrne, 2005)

The ponderings are from a blog contributor writing about the iPod specifically, but the comments really do apply to any PDMD. The combination of massive portable memory space and the random/shuffle function is something wholly new that sets these devices apart from their Walkman-type predecessors. And according to French semiotician Roland Barthes, listening (from an anthropological viewpoint) is still "the very sense of space and time" (Barthes 1985, p.246). So by turning on their devices and walking through the world PDMD users are able to, at least to some extent, turn off what surrounds them. It has to be acknowledged that "the radio was the first sound wall, enclosing the individual with the familiar and excluding the enemy" (Schafer 1977, p.93), but there is something about the individual mobile listening experience that should prompt us to consider this new sonic-cyborg as essentially different in kind from that enabled by the stationary radio, record player, or tape deck of yore.

TEMPORAL AND SUBJECTIVE REFRAMING

For me the question centers on the nature of attention. I grew up hearing that *time flies when you're having fun*. This is the expression, but I've always thought that it isn't quite right. It seems more accurate to say: *time flies when you're engaged, when you're paying attention to something other than time*—this is, of course, why a watched pot never boils. And in most cases we program our PDMDs with music that captures our attention with lyrics and chord progressions that we know or can reliably predict. Even when we engage the shuffle/random play function on our devices, we most often have a decent gist of the music we have loaded onto them.

To be sure, there is often something predictable, something anticipatory and therefore something participatory to our listening these days. Margaret, a nineteen-year-old college student puts it this way: "My iPod is a way that I can have music match accordingly to my mood. I find music to fit my situation, and my life. I have endless songs available at my fingertips to listen to while I am walking somewhere, waiting somewhere, exercising somewhere, or even writing a paper and doing homework somewhere." Heather, continues: "I like to use the shuffle feature on my commute, when I'm not really in the mood for any particular music and just want the solace and distraction of music in general. However, all of my songs are divided into playlists either by artist or activity, meaning I have playlists with labels like: "gym," "mello," "angry," "thoughtful" . . . those playlists change from time to time but they have a general 'mood' to them."

This playing with attention and time through programmed soundscapes alters the agency equation. And this is another common theme in our Digination today. If the idea of relinquishing control was anathema to urban youth a decade ago, today it's what it's all about. With the random function, or more recently with *Pandora* program that provides an AI-based intuitive music compilation, we can experience a phasing in and out of the feeling of control—or agency—that has, apparently, been so crucial to that mythic American sensibility to this point.

Within songs that are known we can still retain a sense of control however. In all the randomness that is possible with these digital devices I think about the predictability that is built into the standard twelve-bar blues I have so much of on my machine. Or the comfort I find in Bach's *Prelude in C*. The predictability, the stability, and just pure sense of serenity that accompany these sounds can ease a tense drive into town or a hurried shuffle through a crowded train station.

But we should then juxtapose this with what can occur between songs on random play, for that one or two seconds as we are waiting to learn where the music is going next—and where it is taking us. This is when users often describe feeling on hold, suspended, or in transition. One respondent explains: "I do hold sacred the sanctity of the album order. I certainly still listen to them the way they were intended, I respect the artists I listen to enough to trust their decisions. But I've loosened my stranglehold since getting my iPod. The control, I think, is often not as satisfying as not knowing what comes next" (Tom). There is a liminal aspect to this kind of listening. For some it can make the difference between a mundane train ride into the city on the one hand, and a private adventure into the city backed up by a favorite band on the other.

"I use it when I run. I make playlists of the loudest, angriest, most high-energy songs I have and use it to help me run. It can sometimes keep me going when I feel like stopping." I use *Rage Against the Machine* for that usually" (Brian). Many respondents talked about a kind of relinquishing of agency to the machine, some were more anthropomorphic in their descriptions than others. "If I believed it was psychic, that would at least give me control over my own moods. But what actually happens is the computer program which controls the random play function picks a song out of the over 1700 I have stored in its tiny hard drive, and plays it. Consequently, my mood swings from happy to sad. Or whatever. So I'd say my emotional control is switched over to a slick little machine" (Tom). Margaret then puts it all in a nutshell: "my shuffle sometimes makes me into an emotional mannequin."

I want to end this chapter with a series of thoughts from these PDMD users. Their insights seem to cut to the chase regarding the way many of us are intertwining with more than just those thin white headphone cords plugged into our heads at one end, and our machines at the other.

> What first drove me to the life of random iPod listening was the convenience of it, coupled with the fact that I had a half-hour commute in both directions each day on the train. I began listening to whole albums, as always, but soon began skipping through tracks, then only listening to one track before changing my mind and switching to a new album. It was easy enough to scroll between artists, but I wanted my T-time to be peaceful, without decisions. The great thing about iPods, I quickly learned, was that it lent itself to easy shuffling, it practically begged to be used as a tool for random song play. It seemed I'd have to adjust to it. Weirdly enough, I found I actually enjoyed the freedom it gave me. I could turn it on and leave it on. If I didn't fancy a particular choice, I hit the forward button on my remote and, presto, a new artist, a new song, and a totally different me. (Brian)

And another listener fills in the details:

> When the volume is loud enough, my entire world changes. People become characters in music videos, walking to the beats playing in my head; couples on seats turn into jilted lovers turning away from each other; old men are given melancholy histories written across their sad faces. Life simply becomes more interesting when it has an unplanned soundtrack. It suddenly has another dimension, an alternate reality. Sound is at times more powerful than vision you know, because the show continues even when I close my eyes and lean my head against the window of the train. It's unstoppable. (Peter)

I detected an emergent theme in many of these stories and descriptions, one that reminded me of the way some of my Mohawk friends seemed to relate to their computers running e-mail. There is something anthropomorphic, and symbiotic going on that extends beyond simply naming our machines or talking to them when they do what we want them to—or when they malfunction. Consider Brian's take: "My iPod proved that it is the smartest person I know, and that it has a sick sense of humor and impeccable taste in women." Brian is a college senior who regularly plugged in during the walk to and from his marketing internship in downtown Boston. Describing the uncanny way the crescendos and drum rolls of his favorite songs often seemed to coincide with notable occurrences going on around him, the remark about women concerned several occasions when he was rounding a corner as the music was swelling to see "the woman of his dreams" appear before him on the other side. This kind of fantasy life was also a common theme that emerged during my investigations.

For Brian, the 80s British rock group James sums it up pretty well with their lyrics to the song *Born of Frustration:* "I'm living in the weirdest dream, where nothing is the way it seems, where no one's who they need to be, where nothing seems that real to me . . . when the world is spinning endlessly, we're clinging to our own beliefs . . ." We'll see next, in the chapter devoted to a different kind of listening enabled by our PDMDs, how certain features of the human voice, the voices of others, does much to help color and even reconstitute *one's beliefs.*

NOTES

1. For an extended and excellent discussion of Debord's and De Certeau's theories of city life, see the chapter entitled "Pavements and Paths" in John Urry's *Mobilities*, 2007).

Chapter 8

Podcast Consumption

From Sound Track to Narrative Track

"It is the sheer quantity of information which has alienated us from political and social reality. The large city isolates the individual citizen, but the multi-cultural perspectives of the press have isolated the human spirit itself from any milieu."

—McLuhan, *New Media as Political Forms* (1955)

In the previous chapter we considered some popular uses and interpretations of the personal digital music device (PDMD) including Apple's ubiquitous iPod. I described some of the ways in which our everyday soundscapes are themselves products of complex cultural processes. From the hum of cars on the street, to the songs of birds in the park and the chatter on our cell phones, the sounds of contemporary life have been subjected to a variety of institutional, political, artistic, and technological manipulations. Here I'll extend that exploration of our emerging aural ecologies by investigating the practice of podcast listening. Podcasts move us beyond music listening to the consumption of talk, conversation, and other kinds of oral discourse as re-mediated through digital recording technology. These recordings of human voices consumed in a variety of contexts are, on the face of it, personal sound experiences. But they are often highly persuasive socio-political events as well.

Several medium theorists can again help us make sense of podcasting and situate this important piece of our Digination. Marshall McLuhan's distinction between percepts and concepts, Walter Ong's notion of *secondary orality*, Federman's *publicy* concept, and Donald Horton and Richard Wohl's characterization of the *parasocial relationship* together provide insight into the social, political and phenomenological significance of real-world podcast consumption.

The growing popularity of this new media form is, I think, due primarily to the power of the human voice. Podcasts seem to be the latest manifestation of Walter Ong's *secondary orality* concept. In the tradition of the beat poet, the DJ, and, most recently, the blogger, podcasting recasts the personal experience as well as the actual perspectives and preferences accessible to people on-the-fly via digital storage and playback of audio content. Given that we've already considered the personal musical experience as a particular form of sonic consumption, I'll focus here on the peculiarities of spoken language reproduced and replayed through digital devices on the go.

Like blogs, podcasts blur our traditional understandings of "public" and "private" thought and experience. In the chapter discussing the social and political significance of blogs and blogging I invoked a recently coined concept. We'll recall that *publicy* is Mark Federman's term for privacy that occurs under the intense acceleration of instantaneous communications. As Federman points out, "our notion of privacy was created as an artifact of literacy—silent reading led to private interpretation of ideas that lead to private thoughts that led to privacy" (2003). But while the podcast may be the latest and perhaps purest form of publicy, is publicy itself really anything new, or is it just a new form of orality (i.e., secondary orality) that recasts us all as witness/participants to the chants and decrees of the iconoclast, village elder, and cultural mystic? These are intriguing questions that hold great import for the shape and feel of our emerging, digitally enabled social and political landscapes.

A BRIEF HISTORY OF PODCASTING
AND PODCAST CONSUMPTION

MP3 audio files (most featuring music) have been swapped in person, and online for more than a decade. *Podcasts* (audio files of varying types often featuring news and talk) were first deployed on the Internet in mid-2004. Podcast consumption has similarities to both the traditional radio broadcast listener's experience, and the tape-recorded radio show played back on a stationary tape player or mobile Walkman. I located numerous "idiot guides" and manuals online and in printed form for the budding podcaster (i.e., podcast producer), and I think the fast adoption rate of the podcast by Internet content providers and users seems to be due primarily to the ease with which these files can now be manipulated. The "one-click-and-you-get-what-you-want" formula has been found.

As more news and media services turn to the pay-per model in their efforts to contend with a series of challenging industrial and economic trends, content producers and consumers are realizing that podcasts can be a cost effective alternative to maintaining subscriptions to various online and satellite services. National Public Radio, for instance, is moving vigorously in this direction, with programming content available via podcast, Facebook, and Twitter. For the media consumer specifically, the advent of large memory capacity, long-life batteries, file archiving on the Internet, and digital encoding make the content and the experience of content obtained through a podcast quite distinct from that of the traditional radio broadcast—whether that be a typical news segment, a celebrity DJ's themed music-banter hour, a call-in interview on a political talk show, or a church sermon.

However, the term *podcast* is misleading, as neither creating nor listening to podcasts requires an iPod, or any portable device for that matter. The name has been a marketing coup for Apple Computer Corporation. Whether the name achieves the eponymic status of *Coke* or *Kleenex*, or begins to slip as more consumers utilize these sound files in a wide variety of non-Apple products, the outcome will not alter the ever-increasing ease by which DSL or T1+ connections, super-fast processor speeds, and the digital MP3/4 formats allow the downloading of compact sound files that can be played back and/or edited on even the most basic audio player programs and devices. The podcast data file encodes a relatively high-fidelity audio signal that is ideal for voice reproduction.

Technically, the term *podcasting* refers to the production and/or preparation of digital sound files posted on the Internet—not their consumption. No definitive term has yet surfaced that references the growing audience of podcasts in all their stripes (like "radiohead" or TV junkie). "Podpeople" is one suggestion a colleague offered a while back, though the allusion to automatons and body snatchers is certainly less-than-gracious and, most likely, extreme. Unfortunately, the allusion to broadcasting is also misleading as nothing is being broadcast per se. To the contrary, and well beyond the casual tuning-in of a radio or television signal, podcast listeners must consciously point their browser to a particular web site that archives these compressed digital recordings and deliberately choose specific files for download. More recently, "aggregator" programs can be set up that automatically check for and download the desired new content, however the source site, topic, genre, and production time-frame still has to be pre-selected by the media consumer. Some of the conscious deliberation has also eased more recently with the "smartest" aggregators even transferring the latest content to one's mobile listening device immediately upon establishing a hard or wireless connection to a host computer or network.

The majority of podcasts feature talk—not music—so this prompts us to shift our thinking from those features and effects associated with the mobile soundtrack (first popularized with the SONY Walkman) to the mobile narrative track. Similarly, but often without an explicit set of physical referents, we'll see below how the religious sermon, the DJ's commentary, or the political pundit's rant nonetheless performs an attention-directing function. Given the unique features of the human auditory sense, there is a tendency to incorporate or *fold in* what we see, taste, smell, and touch with what we hear. If the podcast originates in sound, it often ends up being a more total or whole sensory experience.

Podcast practices are not limited to portable consumption, but I focus on the mobile phenomenon here for three key reasons: First, listening to podcasts on-the-go is fast becoming a mundane activity in urban centers, especially where many commuters who use public transit or pedestrians navigating the high-rise city-scape report considerable trouble tuning in radio signals for a variety of purposes (such as listening to sports, news or their favorite radio talk shows). Second, listening to portable podcasts raises many intriguing questions concerning a person's subjective understanding of the world as experienced on a minute-by-minute, even second-to-second basis. In this respect, podcasts and the appliances that enable their consumption, are among the latest instantiations of mobile digital communication technology that represent a further alteration of the phenomenological experience of everyday life. Finally, the in situ consumption of podcasts fundamentally alters the meaning and import of certain kinds of content. Especially where clichés and generalities are regularly employed as a key feature of discourse (as is very often the case where religious and political talk is concerned) the *mobilization* of such content has been shown to reorient the listener to the world and the world to the listener, prompting (internal) memory and (external) layout to function together as props and foils for the often detailed yet punctuated discourse that typifies the podcast.

THE PHENOMENOLOGY OF THE MOBILE PODCAST LISTENER

In the previous chapter I described some of the key features of the phenomenological approach as applied to the new experienced realities—or *lifeworlds*—bound up in the creation of the mobile soundtrack. Recall that phenomenology in general and the lifeworld concept in particular concern the way reality itself consists of objects and occurrences as they are immediately perceived or understood in human consciousness. Many research participants who ended up contributing to this study reported marked perceptual shifts while listening to podcasts of their favorite personalities—especially the mobile experience employing dual headphones.

The formidable power and mystique of the disembodied human voice was immediately obvious with the first successful telephone conversation between Bell and Watson in 1876. However, it was Reginald Fessenden's 1906 Christmas Eve radio services intended for merchant mariners and the crews of ships at sea from Brant Rock, Massachusetts, that firmly demonstrated the emotional and persuasive capacities of voice-over-distance. Fessenden talked informally, sang along with his wife while playing the violin, and read passages from the Bible for his listeners just off the coast . . . and perhaps beyond.

Even though Fessenden's financial backers did not express any interest in vocal communication (or even the transmission of music for that matter), his work and play with early radio apparatuses set the stage for significant developments that would follow from the electrification of the human voice. Some of these developments include the broadcasting-qua-publicizing of private thoughts; concrete, real-time referencing of local and distant events; the direct-address of audiences; and the subsequent development of compelling, and often enduring *parasocial relationships* between audience members and their favorite media personalities.

> One of the striking characteristics of the new mass media—radio, television, and the movies—is that they give the illusion of face-to-face relationship with the performer. The conditions of response to the performer are analogous to those in a primary group. The most remote and illustrious men are met *as if they* were in the circle of one's peers; the same is true of a character in a story who comes to life in these media in an especially vivid and arresting way. We propose to call this seeming face-to-face relationship between spectator and performer a *para-social relationship* (Horton and Wohl, 1954; 215).

The podcast, and particularly the podcast listened to on the move, may be part of an evolution in parasocial phenomena and a fundamentally new form of mediated interpersonal communication. Podcasts enhance the personal feel and all attendant psychodynamic effects of Fessenden's primordial radio show. They also add infinite mobility and manipulability to the equation. Individuals participating in the research study informing this chapter listened to podcasts of religious sermons, news magazines, political pundits, and music DJs. They often described a kind of organic connection and an enveloping, holistic involvement as part of the experience. Marshall McLuhan's description of a media-induced blending of the senses, or *synesthesia*, helps to account for some of this. Such an effect can occur because sound is a naturally incorporating or inclusive perceptual experience. Hearing, by extension, binds us to the world, often making it difficult to discern where we are in relation to the thing making the sound. Vision, on the other hand, is a highly focusable, abstracting, even distancing sense that puts the seer outside of that which is perceived. This is a key distinction between sound and vision. Sonic percepts simply are not parsed out by the human ear in the way visual percepts can be isolated by the eyes. This is easily understood every time I listen to Bach's Brandenburg concerto no. 2 (1st movement). The oboe's acrobatic dance can't be fixed in time but it can be tracked and separated from the other audible strains without too much trouble. However, the basic quality of the instrument borrows some of its essence from the relation it develops with the mix of strings, harpsichord, horn, and flute. In the process of consciously attending to just one instrument, the experience of the "surround" is often literally muted or set off-color. As listeners, we can't help but naturally hear the oboe as a living, organic, emergent part of the whole ensemble.

For similar reasons, the mobile headphoned podcast listener of the human voice brings Ong's sonic incorporation thesis into its own. Even a short (two or three minute) podcast can uniquely focus the attention, mnemonic, and persuasive functions of the church sermon, the radio talkshow, or the nightly television news hour such that a holistic experience can be had. There is something ironic and at first unexpected about the podcast as a popular form of media consumption. While growing in popularity, the podcast seems to be a rare manifestation in a very broad array of media options these days that are disproportionately skewed toward visual representation. The podcast itself represents a direct turning away from the visual/analytic mode, back to what McLuhan called the "tactile embrace" of the oral/aural. While the idea of secondary orality (as a re-emergence and reorientation of many of the primary oral cultural ways of being) has been crystallizing since the mid-point of the twentieth century, the sense of presence—and the immediate sense of the present—that is encoded in the human voice is further enhanced when the listener can center sonic emissions around the head with low-cost, high-fidelity miniature speakers abutted to or affixed inside the ear canal. With this equipment, we'll see how the modern version of a soothsayer or village elder is functionally empowered because they are aesthetically enhanced. Some of my respondents described it as even more compelling than the televised audio-visual representation.

Indeed, the auditory sense is particularly open to active/additive referencing, association, and expansion through the inputs of memory, the "mind's eye," and the other sensory modalities in the reality-building process. Hearing may be rivaled only by the olfactory sense in this regard.

Psychologist JJ Gibson's (1979) general theory of perception "as an achievement of the perceiver" further bolsters this view. According to Gibson, animal perception is not a passive phenomenon. Rather, perception is always a participatory, active, ongoing process that is only partially (perhaps minimally) under the conscious control of the perceiver. While the stationary media user is certainly free to imagine and associate at will in his or her location-based situation, we are not prompted, cued, or, perhaps, distracted to the same degree as the mobile case (walking, running, in automobiles, or on bikes, boats, buses, trains, and planes).

Moving through physical space, the listener can't help but bring the world they confront into the unfolding monologues and conversations cached on their digital audio device. As mentioned above, in the mobile case the physical environment functions as a kind of perceptual set or backdrop, thereby recasting the recording as a form of discourse, requiring the user to integrate and match up the visual percepts emerging along the way with the audio. There is something undeniable and tenacious about this whole process, and it creates a new hybrid kind of media experience. And all of this should prompt us to rethink McLuhan's traditional characterization of audio media as engendering "hot" or low participation experiences. To be sure, his hot-cool typology becomes a bit confining in this new media context. There are some analogous media worth mentioning here to see how and why this might be the case today.

With hard-wired/location-based telephony moving steadily toward obsolescence, the mobile phone experience edges closer to the cool end of McLuhan's spectrum. Mobile telephony is becoming cooler/more participatory and a generally more involving and distracting activity because we are not allowed to focus as easily on simply what we hear. For similar reasons, what appears to be the most popular (i.e., mobile) podcast experience in my investigations is also a more involving enterprise. The mobile podcast experience is perhaps even more participatory than the mobile radio experience for reasons already detailed in a previous chapter. The suggestion was that humans are not passive, but active information processors. The idea finds support in the work of Gibson (1979) and Avery and McCain (1982), who described early on what we know today to be a sort of truism: the human sensory apparatus is active or homeokinetic.

Humans are, in other words, essentially associative, searching, interpretive systems that try to sustain equilibrium (physical, psychological, etc.) through time. If there is any consistent component to "human nature," it might very well be that urge, that felt-need to find meaning in the world that corresponds to one's current understanding of it. We are always looking for examples, illustrations, and connections—especially if they are presented in relation to a set of ideas one already believes or at least holds some stock in. This is the "homeokinetic" tendency in a nutshell.

Granted, based upon the data collected from the small number of respondents interviewed for this inquiry, no firm conclusions can reliably be drawn regarding the relation and correspondence of podcasted content to the

philosophical, religious, or political predispositions of these podcast consumers. That part of the question of consumption patterns will have to be deferred
to a future study. However, we can make a preliminary set of claims now
regarding the likelihood of using certain content for lifeworld-building or at
least "world-view-building" purposes (for lack of a better term).

Again, the argument here is that human beings are incessant interpreters.
We do not passively stand in a wash of stimuli. Instead, we tend to try to
incorporate, in some meaningful way, the dynamic perceptual flux going
on all around us. And when individuals select particular podcasts, what
often informs and attends that media selection is a heightened perception
of personal relevance regarding the content contained therein. Even in the
face of our rapidly expanding universe of media experiences, all of that
experience nonetheless remains, at bottom, "local." Short of some cognitive
impairment, psychedelic drug use, and barring for the moment any potential
for astral projection, human beings are enmeshed in a highly localized state
of consciousness. That is to say, despite the appearances of multi-mediation,
we remain physically embodied beings. Despite the earnest efforts of post-
humanists like Stelarc, we still can't shed our human skins. And so aside from
periodic lapses of Internet-induced stupor, we can't help but experience the
world in-context, or *immediately* so to speak. If modern media experiences
do a lot to make the body transparent to consciousness, the body, or what
I like to call the brain-body continues to do a lot of perceptual work for us
(was it Freud or Deleuze, or both, who liked to say that the unconscious sees
at infinite speed?).

To understand why this might be so, we have to first re-conceptualize our
understanding of the brain-body as medium. What's more media must always
be understood as *media-in-use*. Otherwise, we are not really talking about
media at all. Just as *radio* is never just "what's on" the radio, the *podcast*
is never simply the sound file. Put another way, there is no media without
a media consumer (reader/viewer/listener/experiencer). To have meaning,
these must all eventually or ultimately be instances of symbolic content
consumed and "filtered" through the particular biases of someone's personal
history, their medium of choice, the particular context in which they dwell
and the biological organism that comprises who and what they are. This is a
simple but subtle aspect of media use that tends to be overlooked by many
observers—casual or otherwise. Let's consider a few examples to clarify this
very important point.

With the first inscriptions on stone tablets, papyrus, and other early portable media, the content a medium carried became unbound from its original
context. As the content of a portable medium is consumed, that content is
also unbound, in a sense, from the medium that carries it precisely due to

its carrying capacity. And yet humans have always used mediated content in particular places. That is to say, our media consumption is always embodied, as well as embedded within particular spatial-temporal contexts. So while portable media are, by design, not bound to particular places, the meaning(s) we glean from our media use has always nonetheless been context-dependent. Just as a particular book read in the quiet comfort of one's home study is not the same book read on a busy subway car, a podcast of Rush Limbaugh's radio show listened to while dozing off in bed is not the same as Limbaugh listened to on the frantic walk from the subway to the office. These are uniquely dynamic processes of active mediated perception and experience that are always our own private experiences. They are inherently *personal.*

The preceding suggests why ostensibly "aural" media experiences have to be considered with particular care and nuance. I want to suggest that a relatively poor acuteness (in terms of reception and gain) best characterizes what is really just a vestigial sense of hearing in human beings today. We are still equipped with multiple sense modalities as our tribal ancestors were, though we are now heavily biased toward the visual sense and the stimuli that sensory apparatus is tuned to. It is very likely, all other things being equal, that modern humans do not hear or see the same way ancient humans did. Yet McLuhan suggests that the *tribal echoland* endures inasmuch as hearing maintains "no margins" in the experience of the modern human. Indeed, due to the physics involved, sound tends to envelop us. There is no perceivable "edge" or border to auditory sensations as there is in a visual display (whether that display is bracketed by our brow, cheek, and peripheral views, or in the four edges of a video screen). We are always, in a sense, at the center of sound. But the aural participant is not so much the center of what is happening sonically, as they are part of an undifferentiated whole—an always less-than-clearly-defined experience. Put another way, hearing is more associated with abstract concepts, as opposed to the relatively concrete percepts of vision. This is one of those differences that make a difference and so will be elaborated upon a bit more here.

ONG AND SECONDARY ORALITY

In *Orality and Literacy* (1982) Walter Ong built upon McLuhan's characterization of the perceptual habits fostered by oral, written, and electronic media cultures. Ong went on to describe the cultural milieu of contemporary electronic media as one of "secondary orality." Primary orality, Ong suggested, cultivated and sustained a special kind of communicative experience that began to fade and lose traction on the human psyche with the advent of the phonetic alphabet about five thousand years ago. That way of being lost further traction with the broad dissemination of the printed word following Gutenberg's moveable type press in the mid-fifteenth century. A progressive internalization and privatization of the phenomenal lifeworld continued through the nineteenth century until instantaneous electronic communication was made possible with the invention of the telegraph at century's end. In the third chapter of *Orality and Literacy* Ong lists a number of characteristics that define the way people in a primary oral culture think, perceive, and express themselves:

1. Expression is additive rather than subordinative.
2. It is aggregative rather than analytic (or separating)
3. It tends to be redundant or "copious."
4. There is a tendency for it to be conservative.
5. Out of necessity, thought is conceptualized and then expressed with relatively close reference to the human lifeworld.
6. Expression is agonistically toned.
7. It is empathetic and participatory rather than objectively distanced.
8. It is Homeostatic.
9. It is situational rather than abstract (i.e., percept over concept).

Ong, like McLuhan, drew analogies between these features of primary orality and our modern forms of mediated interpersonal communication, including various forms of "intimacy-at-a-distance" (Horton and Wohl, 1954). Much in the way a campfire story told in the right place and time commands rapt attention and fosters intense participation, the agonistic modes of address so typical of modern radio or television news programs (conveyed through variances in volume, tone, gesture, eye contact, and emotionally laden intonation) intentionally prompt and then often appear to correspond in kind to reactions in the audience. We see this in David Letterman's stoic gaze after dropping an obscure one-liner or non-sequitur, or Jon Stewart's verbal hedging upon repeating some circumlocution he found in the news, or Rush Limbaugh's hearty guffaw after offering one of his over-wrought interpretations of the President's statement to the press.

PUBLICY AS SECONDARY ORALITY

Fessenden's ability 100 years ago to publicize the private in real-time in a manner never before possible has more recently been described as a form of *publicy*: "privacy that occurs under the intense acceleration of instantaneous communications, . . . [publicy is] an artifact of literacy . . . silent reading led to private interpretation of ideas that lead to private thoughts that led to privacy" Federman (2003). The essence of publicy as a side effect of secondary orality (itself derivative of primary orality) is captured in the following Playboy Magazine interview featuring McLuhan published in 1969. The instance is worth noting because, as the interviewer makes clear in a short preface to the interview itself, this was one of the rare occasions when McLuhan's penchant for oblique, meandering explanations was largely held in check. His thoughts were focused and clarified in the give-and-take process of face-to-face dialogue. As is the case with many of the podcasts downloaded and analyzed for this study, the inclusion of an interlocutor is central in creating traction and a kind of resonance in the audience. In an effort to get to the bottom of a point McLuhan just made regarding several distinctions between oral, print, and electronic culture, the operative question asked by the interviewer is simply put: *What do you mean by "acoustic space?"* McLuhan then replies in what appears below to be a lengthy diatribe, but is, in fact, an uncharacteristically concise response to this sort of question:

I mean space that has no center and no margin, unlike strictly visual space, which is an extension and intensification of the eye. Acoustic space is organic and integral, perceived through the simultaneous interplay of all the senses; whereas "rational" or pictorial space is uniform, sequential, and continuous and creates a closed world with none of the rich resonance of the tribal echoland . . . The man of the tribal world led a complex, kaleidoscopic life precisely because the ear, unlike the eye, cannot be focused and is synaesthetic rather than analytical and linear . . . By their dependence on the spoken word for information, people were drawn together into a tribal mesh; and since the spoken word is more emotionally laden than the written—conveying by intonation such rich emotions as anger, joy, sorrow, fear—tribal man was more spontaneous and passionately volatile. Audile-tactile tribal man partook of the collective unconscious, lived in a magical integral world patterned by myth and ritual, its values divine and unchallenged, whereas literate or visual man creates an environment that is strongly fragmented, individualistic, explicit, logical, specialized and detached.

The present analysis is an effort to illustrate several important points made by McLuhan in this key section of the interview. Even this instance of *McLuhan being clear and distinct* remains fairly opaque by most standards, so I'll do what I can to unpack the gist in what follows. Most notable, perhaps, is the "organic and integral" experience of the world resulting from the "simultaneous interplay of all the senses" that itself stems from being immersed in an acoustic milieu. The experiential wholeness and synesthesia that incorporates the headphone-clad podcast listener takes an otherwise diffuse listening experience and puts them dead-center in the middle of it all.

A key distinction between the primary and secondary oral situation is a heightened degree of self-consciousness (attended by a kind of *self-centeredness*) in the latter case. Whereas an individual sense-of-self (the phenomenological aspect of being more or less distinct and separate from the world) is not characteristic of a primary oral experience, such self-centeredness typifies the secondary-oral situation. Regarding this, Meyrowitz' (2004) observed that "all experience is local," and this applies well to the experience of being connected to a digital mobile listening apparatus. Words or music, in fact, the same applies. We seem to become the nexus of all that is occurring. Again, I'll argue that this has much to do with our being embodied beings having physical bodies of a particular configuration. And now we'll see below how the use of stereo headphones is key to enhancing this localizing effect Meyrowitz describes.

The idea of a modern experiential situation mimicking that of an "Oral/Tribal" way of being was accurately predicted by McLuhan in *The Gutenberg Galaxy* (1962), where he asserted that "our age translates itself back into the oral and auditory modes because of the electronic pressure of simultaneity." For instance, listening to a human voice on the radio, or watching someone on television often fosters a palpable sense of involvement, of being hailed or addressed personally. I experience this any time I manage to be up late, have nothing to do, or feel like doing nothing, and sit down to watch Letterman or Leno. We all know how these viewing experiences are attended by a powerful sense of *being there*. It can be even more enveloping with words spoken to us through a pair of earphones. McLuhan went so far as to say that radio and television were "tactile" media because of the way they can embrace us, facilitate a dynamic interplay of the senses, and "work us over" in the process through this kind of forced synesthesia. Hearing (and smelling), however, seem to be the most resonant of the synesthetic senses.

The experience by an attentive person of a one-way radio or television signal featuring a human voice can sometimes create the illusion of a two-way communicative exchange to the point of even prompting some active emotional, even physical participation in the recipient. Indeed, the parasocial relationship first described by Horton and Wohl in the mid-twentieth century is a cultural commonplace today, whether particular members of the media audience realize it or not. We all *engage* with mediated personae to varying degrees. But this does not require any kind of acting out or talking-back-to (which also occurs often enough). What's much more common is a subtle collection of cognitive, bodily, and emotional experiences. Extending far beyond the way readers of novels might report sympathy for (or may empathize or identify with) various characters, the experience of reading a book prompts such corporeal engagement far less intensely, and much less often. This is simple enough to reflect upon for verification. We rarely talk back to—let alone yell at—a page of printed words.

Podcasts take secondary orality, publicy, and the power of the parasocial relation to a new level. With the headphones in place, we hear someone (or several people) speaking, quite literally, between our ears. And this adds a certain reality to the phrase "getting inside someone's head." Surely, at one extreme, this is what has led some research participants to describe almost feeling possessed by another. But if apparently extreme, such a sentiment suggests that the experience generated by the mobile podcast listening experience is at least akin to having someone speaking to/with you while walking, sitting, or standing next to you. And yet, the podcast retrieves some of the phenomenological characteristics of reading—with a twist.

I used to regularly assign Shelley's *Frankenstein* as part of a special reading list in an undergraduate media theory course. Invariably, upon finishing the book several students (some of whom had seen multiple adaptations of the tale on the theater stage or screen) report a more vivid experience of the story and a more nuanced understanding of Shelley's vision of autonomous technology. Let's briefly examine reading, along with several other modes of media consumption, to compare the experiential outcomes common to each. This will also help demonstrate just how powerful podcasts can be.

While it is increasingly difficult these days to locate "regular," un-coerced readers of novels, their descriptions are generally consistent with those of my students who read novels. It seems likely that the popular experience of reading *Frankenstein* when the book was first published in 1831 (eight years prior to the invention of the daguerreotype) would be qualitatively different than the experience of the same book a century later, after the advent of celluloid film technology. Thomas Edison's 1910 silent (and short) production of *Frankenstein* at his studios in the Bronx undoubtedly recast and redirected subsequent readings of the story. But especially with the large-scale cinematic versions of the mid- and late-twentieth century, we began to see the unprecedented power of a visual medium to prefigure the experience encoded in hitherto text-bound tales. The same certainly applies to early radio audiences of *Frankenstein*.[1]

Reading stories can trigger thinking via *mental picture* in a way that seeing stories often short-circuits. And yet, a particular reference or metaphor is sometimes missed when reading, or re-reading a story. Of course, no claim is being made here regarding the accurate representation of any particular truth or underlying reality in a story. Authorial intent is not what we're after here. Rather, it is merely that experiencing stories in different modes exposes us to a different selection of powerful, but often subtle forms of persuasion. Consider the following three ways of experiencing *Frankenstein.*

Shelly challenges her readers to stretch their senses by filling in the details of the familiar and unfamiliar alike. She walks us through a humid English moor; a mad scientist's laboratory; then finally, the stark wasteland of the arctic tundra. While some of this effect naturally depends on the writer's talents, and a particular reader's imagination, the written words, once read, prompt a filling in of the story with all kinds of additional and residual details. We just can't help but be affected by the descriptions. With *Frankenstein* the book, we walk through a perceptual door. If sequentially directed by Shelley's rich imagery, the imagination otherwise goes to work on its own accord. A serial radio version produced in 1938 adhered closely to the plot and spirit of the novel (and lasted nearly three and a half hours). In that instance we are walked through a different door. The unique listener's imagination is allowed to run a bit more freely with the ability to intermix the actors' language and linguistic cues.

But with the radio the imagination is, in part, driven and guided (in 15 minute segments) by the real-time "action" emanating from the speaker. The radio play, in other words, literally sets temporal delimiters on the way someone imagines the unfolding story. During the action, the pacing and patterning of the listener's imaginings, with all of the added detail, must be in step with the temporal pace and sequencing or the enveloping virtual reality breaks down.

The movie director walks us through yet another door, sits us down in a comfortable chair, and bombards us with their much more explicit version (or vision) of things. As in the comparison between different media forms of Shelley's gothic novel, we might also compare Howard Stern's or Sean Hannity's or Glenn Beck's or Bill O'Reilly's television shows with their podcasts or radio shows. Indeed, as the televised debate between JFK and Richard Nixon profoundly illustrated in 1960, these different media selections often result in very different experiences and perceptions.

Like Kennedy and Nixon, both Stern and O'Reilly also have unique (and peculiar enough) physical features and mannerisms. Here too we find that viewers report being distracted from the content of monologues and discussions to the point of even losing track or interest in what is being said. I'd say this fact alone is reason enough for Stern and O'Reilly to limit themselves to the sound signal. It seems reasonable as well to suggest that the success of Rush Limbaugh and Tom Ashbrook as media personalities is largely attributable to their either remaining with, or returning to, the sound-only medium. In these cases, leaving the visual picture to the imagination enhances the power, presence and, by extension, the very potent parasocial dynamic potentially created between the messenger and his message.

It was established in his 1990s heydays that the average Howard Stern fan tuned in for about 60 minutes (half the duration of his morning show) to hear what he's going to say next. This is an impressive statistic in and of itself. Fan allegiance with Stern was second to none. Of course, many of Stern's harshest critics continue to tune their satellite receivers today to hear him rant tirelessly about the latest political conspiracy, corporate media plot levied against him, or the most recent sex-ploits of his favorite *diva de jour*. Still more impressive, and certainly more interesting, is that fact that many Stern-haters themselves reported spending between 90 and 120 minutes listening, and also to hear what he was going to say next (two of Stern's program directors back at NBC who were interviewed in his 1993 biography admitted to this). While no staunch Stern detractors have yet to be interviewed for this project, we can predict that in such cases as this, the opinion-leader function is not being positively met (though one could imagine cases wherein consistent, long-term Stern listenership might sway individuals less resolved in their previous dislike of *all things Howard*. The same surely applies to Keith Olbermann, Glen Beck, and Sean Hannity.

The consumption (and the production) of podcasts might therefore be thought of as a primarily ritualistic form of communication (Carey, 1989). Carey's "ritual view" bolsters the homeostatic theory of media. While Carey did not focus on mediated contexts per se, he did observe that human communication is constitutive far beyond the simple transmission or exchange of information (the "transmission view"). Of course, Carey's two views are not mutually exclusive. Information obviously continues to be transmitted *through* media. What Carey was getting at here is that we have to also conceive of the symbolic exchange at the root of human communication as also, if not primarily, fostering the therapeutic affirmation of some collective worldview. The sustained worldview then functions to affirm both *broad perspectives* (including political, moral, and religious outlooks), as well as *narrow practices* (habits like listening to the radio during the drive home, reading the newspaper or watching TV at the breakfast table, or listening to a favorite podcast in any of these situations).

Listening to podcasts, therefore, may have less to do with obtaining particular bits of information from elsewhere in order to learn something new, and more with participating in a form of catharsis from the overall sense of the here-and-now that is sustained. On this view, podcast consumption provides a kind of cathartic, ceremonial function. McLuhan's image of folks *getting into the morning paper like a hot bath* conjures this same notion. These ideas align closely with some cognitive and neurological arguments about media making waves today. As cognitive philosopher Andy Clark (1998, 2003) has pointed out in discussions of his "brain-body-world" thesis, the human cognitive apparatus may be a much more energetically conservative and broader set of synergies between bodily components and features of the physical and symbolic environment than has hitherto been proposed. Clark describes a collection of "inbuilt synergies" that are exploited to enable the high level of coordination we see in human behavior and thought.

> It increasingly appears that the simple image of a general purpose perceptual system delivering input to a fully independent action system is biologically distortive. Instead, perceptual and action systems work together, in the context of specific tasks, so as to promote adaptive success. Perception and action, on this view, form a deeply interanimated unity (Clark, 1998; p.8).

Avery and McCain offered an early version of this argument with their homeokinetic thesis. And this may explain the tendency in humans embedded in an oral milieu to be "conservative" as McLuhan (and then Ong) suggested. Of course, to act conservatively in this sense does not imply adopting the political disposition opposed to liberalism. While it could include that perspective, the conservatism McLuhan and Ong described concerned the active and ongoing maintenance of a world view: the constant effort to reify a status quo of familiar habits, traditions, and ways of being in (and seeing) the world. If this seems tenable, then the communicative processes fostered by many of our new media forms, including the podcast, may end up being much more ritualistic than informative in both nature and function as James Carey (1989) opined.

THE PRIMACY OF THE PERCEPT

In *New Media as Political Forms* (1955), McLuhan wrote "one is perceptive when something is penetrated, extracting uncommon insight. Perception is enhanced when attuned to the 'secondary' senses, the tactile, olfactory, and acoustic. Only when the senses are at work, can the eye see. Percepts function via the sensory world, not by concept. Percepts are participatory, involved. Percepts feel . . . transmit subliminal energy."[2] The distinction McLuhan made between concept and percept offers us some insight into the shift (back again) from the culture of the written word to that of the spoken. "Concepts in contrast are detached systems that neutralize participation by explaining the world. Concepts distance us from objects by relying on the passivity of the eye" (ibid). The synesthesia that can be had with the podcast experience recombines concepts with the persuasive concreteness of wholly enabled human perception. I want to then argue, somewhat counterintuitively perhaps, that the podcast brings us back to a time before electricity, the ability to record, or even writing.

With the mobile podcast, and more than any other media form perhaps, we tend to see the emergence of a powerful kind of "horoscopic effect" regarding a listener's relation to the content that has been selected. Media use in the digital age (with the media consumer now able to perform "easy ins" and "easy outs" in unprecedented ways) can allow us to systematize the "cherry picking" of data and information. Selective perception often occurs where one seeks to find correspondence between their "theory" or concept of the world, and the extant percepts in the immediate and mediated environment—what amounts to their sensory/corporeal experience of that world. Some of my research participants have indicated that similar processes seem to be at play during their consumption of podcasts. Much work in media studies and media ecological studies in particular recognizes the importance of something we might call the *primacy of the percept* in individuals' subjective experience of the world, as industrialized societies move progressively away from a typographic focus, to that featuring image and sound as the primary modes of symbolic representation. Again, with the latter we find many of the "concrete realist" cognitive and behavioral tendencies common to a primary oral milieu re-emerging, including suspicion, superstitious/mythic thought, spontaneous emotional response, and the like.

Here, as with blogs perhaps, the user becomes (part of) the content, as they find themselves participating as *pod-clad* "news makers" in the reality-building process. Consider the "happenstance" of the daily commute while listening to a favorite podcast. The potential to override a conceptual way of thinking by a more concrete or percept-based mode of thought and awareness is well illustrated in the following statement made by a research participant ruminating on his recent podcast experiences.

> Look, he's right you have to admit it. There is a really long history of protectionism in this country right? And people who care about this country *protect us* by buying local, buying American. That just makes sense. And, I know a lot of liberals, and they don't tend to buy American. I mean just look around. Who's in the Toyota Priuses?"

Notice how the cherry-picking process is in full effect as Sam (a male sales executive in his early forties living in the Jamaica Plain neighborhood of Boston) interprets the observation made by the host of a conservative radio talk show that "liberals don't buy American" as a kind of axiom.

Or consider the following statements made by a church preacher. His podcasted sermon about greed was downloaded by a woman who listened to it via her iPod as she strolled through part of Boston's commercial center. The preacher, obviously a very talented young orator, was saying how "Jesus does not tell us that having money . . . or being skilled at making money is wrong

. . . the problem with the rich man is not that he's wealthy, but greedy." The listener, Erica, a thirty-something office worker, was listening for the second time to a seven minute excerpt of the sermon she had downloaded several days prior. Erica discussed the way she was listening and looking around at the people and things in view while walking through the Downtown Crossing shopping district. She described a powerful kind of resonance in, and new reverence for, the words spoken the previous weekend by the new preacher at her church whose sermon she was unable to attend live.

Wonderfully illustrating Avery and McCain's points regarding the tenacity of the "human information processor," she said, "I can make all kinds of connections . . . I totally see what he's talking about even when I'm not there." The argument I'm putting forward is that Erica now makes "all kinds of connections" precisely *because* she was not there at the live sermon. Mobile podcasts allow the incorporation of any kind of observation into our everyday lives in new ways. It might be this new disconnect between the people and places where utterances are produced and where and by whom they are experienced that adds new breadth, depth, power, and force to the significant citizen or opinion leader. I want to argue that these modern versions of the village elder, mystic, or seer are empowered by the podcast practices now emerging.

The downtown stroller continues: "I never miss the speakers now. And now I notice all these new things during my walk." I asked what kinds of things she noticed and she replied: "well, you know, there are also a lot of pretty unthinking people driving around in really big cars still. I mean, look, my dad has a big car, but a lot of people in the older generations can't shake the belief that bigger means safer. But they're wrong, it's not true." Can she really hear everything in the bustling tumult? When the ambient sounds begin to overwhelm the recorded voice, Erica simply turns up the volume. "Yeah, the resolution is really good. I can hear him take breaths between sentences. He's a great speaker. And it's like he's next to me."

Our mobile parishioner continues: ". . . or like when you see somebody's husband pull up and double park, filling the trunk of their big brand new German sports sedan with shopping bags from Prada and Macy's, you have to wonder just who it is they care about in this life . . ." This response to the greed sermon suggest a subtle set of directives with respect to how to interpret the social environment for someone already potentially predisposed to this outlook. Unfortunately, the mobile podcast used in this way may become a pernicious sort of prescription for perceptual distortion. Indeed, does one *have to wonder* in this way? One certainly *could* wonder this, and might even be accurate in making such assumptions. But as far as intuiting the precise subjective cares and concerns of those unknown persons (albeit with poor parking etiquette) who were finishing up a shopping excursion and who *seem* to fit the linguistic description is, to be sure, a leap of faith. Erica potentially errs here due to her spatial and temporal displacement. Of course, decontextualization is a common by-product of digitization. It also happens to be a common *way of being* in Digination.

Certainly, one can, and probably should, wonder otherwise—that is to say wonder additionally. However, indexical statements and explanations about the world such as these, or even more explicit statements regarding just why the world is the way it is happen to abound on talk-radio and television with entertainment, news/information and the hybrid infotainment genres now coalescing. I'd say the content of so much that is podcasted (and downloaded) today may be more accurately characterized as a form of infotainment and ritualistic self-assurance. This has already been hinted at with Carey's ritual view of communication, and thinking about podcasts as the *gathering of assurances* (as opposed to the intake of information) certainly seems apt.

The role of environments and the binding power of contexts of different forms is a recurrent theme in this book because it is a central feature of Digination. I can offer at least one personal example to illustrate the potential context has to guide both thought and action. It relates directly to the current discussion of religious talk. Until I was a teenager and stopped attending, going to church and listening to the priest's sermon was seldom a savory proposition. My ultimate impressions rooted, I have to think, to the space in which the communication took place. The church our family belonged to was a pretty typical late nineteenth-century gothic cathedral, and, as such, St. John's was a very dark, cold, and musty sort of place. Due in large measure to the physical space, the words being spoken seemed to take on the heavy hue of that immediate environment—drafty, a bit creepy, and just plain loathsome. I am certain all of this contributed to my thinking about and looking for examples that might help illustrate what was being uttered at the altar. I found my referents, but they did not enhance my understanding of the scriptures in the way my pastor might have wished.

I forgot the details of those particular sermons long ago, but had I been able to move about the everyday cityscape while listening to those same words, getting out of context like that may well have made the message more interesting, more powerful, and more positive. It is conceivable that moving religious/sacred content out of its typical spaces and places may be an effective and fruitful way to rekindle waning congregations and build general cohesion. When the world becomes a stage for a virtual pulpit, new kinds of powers are unleashed. The social sphere becomes a backdrop to twenty-first-century morality play writ large where just about anything can happen, and yet still be explained—or explained away.

Many Muslim and Jewish communities are embracing these new technologies in haste. The Catholic Church seems much more resistant so far. Ironically, podcasting sacred language may be what many institutionalized religions have been waiting for, though I do get the sense (given Erica's remarks above) that this might also open the door for new strains of fundamentalism.

Much like a horoscopic/astrological effect, mobile podcast listeners collaborate in the reality process by helping to create a situation where they see something that provides an example for whatever they are listening to, already believing in, or are seeking to somehow prove to themselves. If you want to see it, the proof will appear! Of course, people carrying shopping bags are not necessarily greedy, but the believer can interpret them to be so at that moment. And an unscrupulous parking job may be just that. With "mobile sermons," as opposed to those previously experienced in the church pew (or even the living room via television or radio), chances of the social world giving rise to this kind of "circumstantial evidencing" drastically expand. So what might otherwise be mundane and familiar can come to perturb perception, setting the stage for this kind of effect.

Think about a walk, ride, or drive across town travelled a hundred times or more. It is likely that you notice little about your environment as you proceed quite automatically from point to point. Similar to the way a music/lyric track can motivate new perceptions, the podcasts' "narrative track" often prompts novel views onto, and interpretations of, the otherwise mundane environments we inhabit. Our mobile parishioner's experience described above illustrates this in a profound way.

McLuhan pointed out how "perception is mercurial, comes out of nowhere suddenly, and is instantaneous, boundless, and involving" (ginkopress.com). On the other hand, "[c]oncept-ualization is static, repetitive, detached, and self-enveloping" (ibid). This is why we may have found in the podcast the most ideal medium for twenty-first-century politics. The citizen consumer reacts in a visceral fashion to the agonistic utterances and exchanges of the sermonizer, the iconic DJ, the politician, or pundit.

Are Sam's and Erica's ideas about the social world points of fact? Are they more tenuous observations? Perhaps a collection of bland generalities? We should put aside, for now, questions regarding the rigor behind the sentiments (or the broader intentions) of the media personalities who uttered the statements prompting Sam's ponderings about liberals, or Erica's thoughts and visions regarding greed. The potential misapprehension, misalignment, and *buying-into* on the part of the listener are what make this so striking. And it seems to be occurring on all sides. It may, in fact, be representative of an emerging pattern that constitutes efforts on the part of many new media consumers who try to reconcile the world as experienced, with the picture of the world created and sustained in their head in the midst of their everyday media use. Regarding the mobile podcast phenomenon specifically (with all of the necessary preparations that go into the process), are similar tendencies to "go with the flow" playing out with the growing numbers of people regularly "jacked in"? There are of course, so many exceptions to this tendency toward groupthink. Indeed, we have to admit that certain media personalities may attract a following for reasons that are, perhaps, more counterintuitive, as the Howard Stern phenomenon discussed earlier illustrates. That is, media consumers sometimes just tune in for the "shock and awe" of it all.

PROGRAMMING PREDISPOSITIONS

The agonistic, emotional tones Ong described so well push a focus on percepts. Like the response more typical during religious experiences, our response to modern political discourse (which is increasingly designed to be very concise so as to fit the various news holes of media outlets) is often also an emotional response. Personality over policy, image or perceived character over idea and content lends itself to a "conservative" activity. It fosters ritual forms and functions to maintain particular cultural forms over time—as opposed to merely extending or exchanging information through space. In short, the content that fits most podcasts seems well-suited to maintaining the "choir," but a discontinuous collection of micro-cultural political enclaves is what that new choir may become.

This description of someone seeking to make sense of the world around them might paint the image of a lone agent distinct and separate in a turbulent sea of anonymous others. But this ultimately supports the social interactionist position that each of us is a *collective*; embodied intersubjectivities inexorably connected at both conscious and sub-conscious levels with those around us. This includes actual people in our physical proximity and mediated personalities floating around in our heads). It was discussed in the blog chapter how Mead's (1934) description of the self as a collection of "generalized others" is one of the first formal articulations of this idea. Of course Mead did not mention the possibility for a kind of inter-personal communication taking place even when there are no embodied others in one's physical presence. But there is now broad recognition in the field that sustained parasocial relationships with a wide variety of media personalities creates a situation where, despite being ostensibly "one-way" communicative contexts, we discover new forms of interpersonal communication taking place. And this certainly complicates the issue of how questionable thoughts and behaviors are perpetrated in the name of staking out the *truth* about the world.

There are an almost infinite number of ways, then, that a listener might take direction or interpretive cues from a mediated message—as there are surely an infinite number of ways those messages can be interpreted. Recently in the news: a man in Pittsburgh shot several police officers while raving about the Obama administration's alleged conspiracy to slowly take away every American's right to bear arms. It has been noted on numerous news programs to date, that the shooter had downloaded content from Glenn Beck's hyperkinetic television show, and had been listening to Beck himself rave about this same issue with aplomb. Just as echoes of the "magic bullet" (or "hypodermic") media hypothesis have again been swirling through the typical outlets (as was the case after the Columbine shootings and Oklahoma City bombing), there is, potentially, something a bit more subtle, and I think pernicious potentially at work here.

Media consumers can now perform the equivalent of a detailed "reading" of the *Turner Diaries* during several walks through the middle of town. One can only guess what sort of props would crop up there. Given this, the rants and decrees of our new shamans and seers may be enhanced beyond measure. What's new to this media equation is the way these platforms carry and, in so doing, re-constitute their messages. Now people can walk down the street alongside Tom Ashbrook or Glen Beck. With podcasts, the world essentially functions as a stage for our predispositions (often aligned with those of our narrator).[3] But unlike the predispositions of our ancient elders—who for mnemonic reasons had to employ aphorisms, and broad generalities, and speak in circles and rhyme—portable recording formats like the podcast free us from the ephemerality of speech. Now all the talk (that matters?) is being systematically archived for download and playback.

If blogs have proven to be more than a passing fad, and as we are seeing less and less discretion on the part of Google, Yahoo, and Bing regarding the difference between news blogs and traditional/mainstream news sources, the blurring of news and opinion is more rampant than ever before. While podcasts need to be posted on websites, they remain, for now, at the discretion of the downloader. They are not subject, like blogs, to the popular aggregations of search engines—at least not yet. Given the way podcasts are selected and consumed it is likely that, for "the choir" at least, the messenger will be significantly enhanced and empowered.

In 2006 folk/rock legend Bob Dylan began moonlighting as a DJ for XM satellite radio. With a receiver module anyone, anywhere, can tune in to hear Dylan take on the persona of an old-fashioned jockey "spinning" those favorite tunes in keeping with his weekly theme. Satellite subscribers did not remain Dylan's excusive audience for long. Bootlegged podcasts of the *Bard's* shows were made available online within days of his first engagement.

While podcasts are not limited to linguistic streams, they tend to be filled with talk. Even if Dylan's satellite radio shows are more music than speech, they are worth mentioning here because for so many people it is the moments between the music, when he makes reference to the contemporary significance of the tune, or offers up an acerbic quip about some recent political folly, that things really begin to resonate in his audience. In these cases, Dylan becomes the father figure, the village elder, the wise man-qua-cultural interpreter for a broad listenership (purported to be in the millions across the country and around the world). I've never been anything more than a casual listener of Dylan's music myself, yet I have to admit that there is something quite magical, and certainly mystical, about his intimate-sounding yet very public ponderings. It is publicy on high. The power of the parasocial effect, and its ability to activate his audience is palpable. Now Dylan can be in your head, or walking next to you at last. Dylan, Glen Beck, Tom Ashbrook, and the religious sermonizers: all serving the same cultural/social/ritualistic function. This is different from the legendary newscasters like Cronkite and Brokaw, who we presumed were giving us the news, and not their interpretation of events. Is the podcast a new medium of hero worship? Indeed, much seems lost in the visual-only mode; in the generalizations of those "thousand words." Listening to Dylan, however, folks get the message—or a message—a bit more directly perhaps. Like the availability of the museum audio tour, the world is interpreted for us. But while resonance with the walking narrative is all but assured as my kid is guided through the *Met*, what happens to wonder?

In the same way the examples and references in the podcasts discussed here can prompt the listener to resonate with "anger, joy, sorrow, and fear" (McLuhan, 1969), these diatribes—these externalized screeds on the move—can push people to action. And this prompts one more thing from the high priest of pop cult: in response to a question from his interviewer as to whether McLuhan was claiming, with our continued move into the world of electronic media, if there will be no taboos in the world tribal society envisioned, he suggests otherwise.

No, I'm not saying that, and I'm not claiming that freedom will be absolute—merely that it will be less restricted than your question implies. The world tribe will be essentially conservative, it's true, like all iconic and inclusive societies; a mythic environment lives beyond time and space and thus generates little radical social change. All technology becomes part of a shared ritual that the tribe desperately strives to keep stabilized and permanent; by its very nature, an oral-tribal society—such as pharaonic Egypt—is far more stable and enduring than any fragmented visual society. The oral and auditory tribal society is patterned by acoustic space, a total and simultaneous field of relations alien to the visual world, in which points of view and goals make social change an inevitable and constant by-product. An electrically imploded tribal society discards the linear forward-motion of "progress." We can see in our own time how, as we begin to react in depth to the challenges of the global village, we all become reactionaries (ibid).

So the oral/aural co-opts the visual. As David Hume opined, *reason is slave to the passions*, and in this new world of sound and vision made possible with the podcast, the passions seem again to be pulling rank.

In Chapters 4 and 5 (Blogs and Internet News in particular) I suggested that the theory of the Internet is the theory of Digination. I said this because it blows open the doors of perception and experience, flattens an artificially narrow set of conversations about the world, and allows bogus and bona fide dialectical interaction to occur between members of vastly different-minded interpretive communities. In this and the preceding chapter, we discover how portable digital audio media and podcasts in particular might be the new media forms that reify McLuhan's oral tribe and Carey's "ritual view" of communication. Examples to this effect abound but I'll offer just a couple here in closing.

Jerry is a married father who describes himself as a "demonstrably not-knee-jerk partisan wing nut, but rather a thoughtful and pragmatic conserva-tive." Jerry mines the online multi-media presence of the National Review and Weekly Standard. Or take Eve, a single mother who refers to herself as an "open-minded, liberal, social progressive." Eve keeps NPR bookmarked and regularly downloads podcasts of *OnPoint, This American Life* and *All Things Considered* for her daily commutes and workout sessions. And here we see how the medium can be the message (or at least predetermine the message).

In other words, not in theory, but in practice, podcasts often ensure a fairly narrow information diet that bolsters what may often be equally narrow views onto the world. This is due, in part, to the podcast's relatively short duration, the intentional way they are selected, and the manner in which they are con-sumed: on the fly, with something in mind, and with a world of things and people serving as the background to the emerging "story."

If McLuhan's newspaper reader at the mid-point of the twentieth century settled into his hot newspaper bath in the morning, today's podcast listener is enveloped in a swirling, multi-jet, hot massage whenever (and wherever) they want. Perhaps more than any other mediated form, podcasts-in-use also tend to blur the distinction between sender and receiver. Even without the ability to call-in to the recorded show, the listener is a cyborg—part medium, part message. Massaged by and massaging the environment, playing with all the props that crop up on the endless horizon of our new surrogate situations, the user (and everything else) becomes content. And so as McLuhan wondered about electronic media on CBC TV one night in 1960: *What will we do with these new teaching machines?*

NOTES

1. This is all well and good, but do modern cinema goers report being less terrorized by, for example, DeNiro's rendition of the monster (as some of my students insist)? Certainly, unscrupulous casting can inadvertently short-circuit the hopes of the director in capturing the right feeling and flavor in a character.

2. www.gingkopress.com/02-mcl/z_mcluhan-and-the-senses.html

3. And these new "surrogate situations" (Clark, 1997) will provide us with endless fodder.

Chapter 9

Knitting, Napping, and Notebook Computers (and a few other mnemotechnical systems)

"The new caravan . . . more multimedia choices than your home"

—2010 Dodge Caravan advertisement

We seem to be packing more technical features into smaller and smaller spaces as our digital infrastructure proliferates. In additional to offering a lot of potential for all kinds of fun, the curmudgeon might say that minivans equipped with DVD players, 16:9 high-resolution LCD screens and multi-point digital surround sound systems might also take a certain something away from the family outing during Labor Day weekend—not to mention potentially contribute to some serious traffic accidents. Consider in this regard the new *SafetyTec* blind-spot monitoring and rear cross path detection systems now standard in the caravan's well-appointed cousin, the *Chrysler Town and Country*. Lexus, Volkswagen, and most of the auto manufacturers around the world are getting on board with these "upgrades." These are, of course, perceptual technologies, and so we might often forget that whatever behaviors we give up (despite careful disclaimers offered by manufacturers) tend to chip away at our innate ability to make our way through the world. But then, this chapter is not about the latest AI-enabled ambulation systems and other digital applications currently being embedded in automobile technology. Below, we'll consider the make-up of some of our intentional thinking and learning spaces.

Indeed, while there may be some serious problems of application in certain academic settings, things like Wi-Fi technology continue to be installed throughout college and university campuses at a feverish pace. Public libraries, coffee shops, and state parks are some of the additional venues now being so enabled, but a number of intriguing questions remain about the wisdom of a process that seems, in practice, to be essentially unbounded.

I've sat on various technology committees at three institutions of higher learning, and from what I have gathered in conversation with other academics and administrators who have done the same, there is often little or no real discretion involved in our application of digital communication technology in colleges and universities. The "newer = better" formula does not get much push-back in most quarters. But here again I'll speak a bit about my own experiences—and interject a few of my own questions—regarding this phenomenon. We can begin by considering some fairly mundane questions. For instance, how attentive to course content are students who have hi-band access to the Internet at their fingertips? What does multitasking presumably related to the day's lesson actually look like in real-time? And, how efficient and effective is this process with respect to enhancing students' experience in the classroom? Some of the earliest data for this chapter were obtained using a *screen-capture* program installed on research participants' machines. That was late 2006-early 2007. Since then I've gathered a lot of additional ethnographic data during observations and interviews in my classrooms, and a wealth of information and insight has also been gleaned from student writing on the topic in daily media journals. But what really put my fascination with cellular and Wi-Fi technology into high gear is this very basic question: *what are people doing with their computers, cell phones, and smartphones while in class, at the dinner table, in their cars, at the office, and during meetings?*

KNITTING AND NAPPING

My original plan for this chapter had nothing to do with a book. At the prompting of a colleague who organized a panel at a national conference, I agreed to collect some initial data and produce a brief research report about the use of laptops in my classes. I wasn't alone. Seven researchers shared their latest inquiries, observations, and experiences regarding laptops in classrooms, libraries, and other academic settings. So in homage to that meeting I'll begin here by discussing a few of the classroom examples. I'll then widen the context of analysis to consider additional observations on computer and mobile phone use in less formal settings. Interwoven here as well is a critical analysis grounded by phenomenological understandings of portable digital media use in a number of "social" contexts.

One of my students in a media theory course a few years back attempted to argue that a light, short sleep during class aided his retention of lecture and discussion material. Despite his laudable attempt to get me to reconsider my classroom policy, I have never tolerated napping during class meetings. Now, this is not to say in any categorical way that a brief nap is ruinous to a student's ability to retain and understand what's going on in class. I can only say that it was not a fruitful strategy for this particular student, and that I have yet to see it work well at all for any student of mine. But, in keeping perhaps with Gardner and Hatch's *multiple intelligences* theory, I do sense that there are a variety of learning styles and habits out there. I am a cautious supporter of methods and means that have the potential to aid students in their retention and understanding of the material covered in my courses. I have, for instance, come to allow knitting. Indeed, I'd love to see more of it at school. Let me suggest why.

I was skeptical for sure, back in the late 1990s, when a young woman came into my Rhetoric and Society class with all her gear on the first day. She politely asked for permission to set up in the back of the room with a basket full of yarn and went at it for seventy minutes, needles blazing. Knitting has by no means ever been a regular occurrence in my fifteen years of teaching, however I eventually became convinced that this perfunctory activity may in fact operate as a focuser and mnemonic aid for certain people. While I've never had a "Rosie Grier" in my class (that is, a male student who knits . . . not to mention an apparently hyper-masculine football player), I've seen how such practices can enhance the ability to attend to and absorb complex material. The subtle movements of the arms; the surgical precision of needle, fingers and hands; all punctuated by a periodic pull on the spool. With the knitter in full motion I get the sense that I am witnessing a kind of biomechanical organism at work. The entire body is so obviously involved. And this of course means the brain too. In such moments, in fact, I get the sense that I am witnessing a larger, extended mind at work. The *knitter-student*: an extended, mnemotechnical being, a low-tech but nonetheless bionic memory and learning system.

In the context of a college classroom, knitting might also be considered a form of occupational therapy, something that potentially even creates a kind of *low frequency cognitive carrier wave* for that rare student who gives it a try. I do know that a couple of my knitters (again, only a handful since the mid-90s) have been people who are prone to distraction, or who are perhaps better auditory and holistic learners. I'm convinced that knitting and other non-cognitively demanding activities can facilitate the task of attending to spoken utterances, and even enable the meaningful retention of relatively large amounts of detailed information offered up in linear format. Part of the neurological story here suggests that knitting and other fairly simple repetitive tasks help trigger the release of GABA (gamma aminobutyric acid) and associated neurotransmitters in the brain, which then assist the work done by serotonin and dopamine to enhance mood, mental acuity, even memory retention. Without getting into the neurochemistry, however, and before considering the phenomenology underlying all of this, I thought it might be useful to simply consider some of the attention-getting strategies employed by other professors that I have either read about or known personally over the past several years.

A 2009 article in the *Chronicle of Higher Education* featured the strategy of Cole Camplese, who teaches educational-technology courses at Pennsylvania State University at University Park. Camplese "prefers to teach in classrooms with two screens—one to project his slides, and another to project a Twitter stream of notes from students. He knows he is inviting distraction—after all, he's essentially asking students to pass notes during class. But he argues that the added layer of communication will make for richer class discussions."

Now, in rather stark contrast to this, a colleague of mine obtained an RF signal generator, or "jammer" that effectively blocks all cell phone usage up to thirty (or so) feet from where it is located at the front of the class-room. The device, which is simply turned on at the start of the meeting, prevents browsing, texting, twittering, and other forms of clandestine cell phone use during class. That particular device did not jam Wi-Fi signals however. Students with laptops were left to their own devices. It is notable that a heavy fine (something on the order of $11,000) is levied jointly by the FCC and FBI upon anyone caught using such a device. And there is good reason for this strict prohibition in certain contexts, including aca-demic ones. The emergency notification systems utilizing text messaging services that were put in place on college campuses around the country after the Virginia Tech massacre are the kind of things that can be disabled by one of these $30 gadgets. So it's certainly a risky undertaking. But it is a risk some are willing to take, as many teachers seem to be nearing the end of their ropes.

I mentioned just two of these very different strategies because they repre-sent diametrically opposed philosophies for dealing with mobile technology use in the classroom. There's no question, I've also been pushed near wit's end by a few students over the past several years who just can't seem to put their devices down. And so yes, I coughed up the $30 to have a federally prohibited device shipped to my doorstep from Korea. I had to see what these things could do. The cell jammers work very well by the way, but the most intriguing thing about them are the responses they generated in some of my students. After an initial, clandestine trial run of my jammer in a basic com-munication course, I was fascinated by some of the comments I got back from students when I revealed that it was, in fact, not a massive "network failure" as I had initially insinuated. Many students were angry at me for cutting out their *iPhone* and *Blackberry* signals, and a handful described experiencing something not unlike a kind of physical, even psychic assault.[1]

Something is definitely going on with our little digital assistants, so let's consider these two strategies just mentioned. On the one hand, with Camplese, we have the liberal theory. His thinking is that all the texting and twittering will prompt more active thought and perhaps periodically create a springboard for more in-depth discussions of the topics at hand. The idea is that students aren't being asked to multitask per se. Instead, the assumption is that they are being equipped to better ride a continuous wave of thought and action. In other words there is, in theory, a single task in the offing: learning the lesson of the day. Let's follow the entire circuit as Camplese might envision it. The circuit begins, somewhat arbitrarily, with the professor's words and gestures, along with any visual accompaniment. These sensory data are then looped through students' perceptual and cognitive apparatus (that may or may not include a knitting rig). The information-turned-knowledge is then turned out through their fingertips to their phones, and on up to the digital "brain-body" . . . and beyond.

In the case of Camplese's Twitter class, students' thoughts, ideas, inklings (delimited to 140 alphanumeric characters at a time) are transmitted up to the Twitter server and back down to the dedicated read-out screen in front of the professor. He then selects whichever comments and questions he thinks best move forward the topic under discussion. According to philosophers Andy Clark and David Chalmers (1998) this is a form of *extended cognition* in which a group of digitally extended humans work together to form a "collective mind" comprised of the professor, his notes, the Twitter system, and his students and their thoughts reflexively produced and mediated through tiny screens held in hand. I'll say more about extended cognition as a movement in cognitive science, the philosophy of mind, and media studies in the final chapter. For now, let's consider a few more of the other classroom strategies now being employed and deployed around the country.

Again, I know of many professors who prohibit peripheral devices of any kind and are betting on the hope that eyes and ears tuned to the front of the room will do the trick of proper comprehension and participation. These folks often require students to turn off all digital appendages at the start of class. Some let the machines run, but adopt an "emergency-only" rule, with the vibration setting enabled. A few others, including me, attempt to strike a bargain with their students by imposing something like a "three-strikes" rule. Still others collect the devices in a basket upon entering the room. "I put mine in the bucket too, as a sign of my commitment and attention," explains one computing professor. She continues, addressing the attendant computer problem: "In the Mac labs, I use ARD (Apple Remote Desktop) and lock the student's screens when I'm giving a lecture. Then they have nothing else to occupy their time."

Other professors I queried on this general question of technology in the classroom expressed opposition to laptop use in particular. Some institutions are growing more relaxed about laptop use, not to mention all of the other devices now at our disposal. A growing number of schools now have policies in place that actually prohibit the instructor from prohibiting laptops in their classes (to allow cases, for example, where a student can produce official documentation regarding an established learning disability). Indeed, many official prohibitions have been deemed of questionable legality of late, as the various idiosyncrasies of the learning disabled gather more legitimacy.

In some cases, then, we seem to have individual minds at work that are not so blended with their devices. The descriptions can range from a nearly classic transmission model of communication where the professor makes "deposits" in the minds of her students, to a more interactive model where real-time discussion moves the hour forward. But we know that *message sent does not mean message received,* and *two monologues do not make a dialogue.* With such caveats in place, there is a reliable pattern emerging in the way people think about the use of various devices, and in the way we tend to think about thinking in general.

In a number of informal interviews on the subject some of my colleagues said things like: "I tell them that I want 100% of their attention while they're in my class," and "I need their minds as well as their bodies . . . so no devices allowed." But one wonders: what is a pen and note pad? Aren't doodles also potentially distracting? What is happening when a student sitting adjacent to a window periodically drifts off with the breeze, or a flock of geese? There is some evidence that doodlers doodle, and day dreamers dream in order to help them focus on the lecture—like my knitters do. The problem is that policies and practices like these might only distract from a more fundamental question—and a much larger issue.

First the question: despite the popularity of signal jammers, official prohibitions, special rules and regulations, or informal agreements, do we ever really get 100% of any student's attention all the time? The answer is clearly no—we never have and never will. None of these strategies are foolproof—at least not toward insuring attention and comprehension. Getting our heads around the larger issue may require something akin to a paradigm shift with respect to the way we think about thinking, mind, and cognition. A personal anecdote should suffice in illustrating why.

Admittedly, I am one of those professors who continues to try and reason with his students. For sure, I give democracy a chance whenever possible, but there is a certain degree of responsibility underlying bona fide collective bargaining. So I wasn't all that surprised recently, when three students in two separate sections of a fundamentals course used up their "strikes" within the first several class meetings.

There is something alarming, even downright disconcerting, about the apparent inability to follow such an ostensibly simple request. We have heard of emotional, psychological, and physical addictions to everything from baby blankets, cell phones, footwear, marijuana, and alcohol. But we may very well be seeing something new in the works. I believe the extended cognition camp is on to something when they talk of a kind of "cognitive addiction" as well. Perhaps this new type should be added to the list if we hope to get a handle on what is happening all around us in reference to thinking, seeing, and being in the world.

What we seem to end up with is a smattering of potentially effective strategies blended with a hefty portion of naïveté in the form of expectations (and ponderings) about the role of technology in our daily lives. One thing is certain—the increasingly ubiquitous use of digital devices today that is emblematic of our Digination has generated a great deal of anxiety for many teachers. It is an anxiety prompted by a felt need, even a frantic expectation that they be listened to, and it seems to be a reasonable enough expectation. But it is also an expectation that becomes more unrealistic (and so idealistic) every day, as the baseline experience of many people includes a near-constant interjection to their flow of consciousness, attention, and awareness by text and image, beeps and buzzes, twitters and tweets. Let me say a bit more, though, about the classroom as a specific context of interaction before we proceed.

A WIDER ATTENTION ISSUE?

There are at least two important observations to be made concerning the purely physical aspect of Wi-Fi'd laptop use in the classroom. These relate to traditional, or "unplugged," and machine-clad students alike. First, there is the line-of-sight question. And here we have a potential benefit that can be a positive gain for both student and teacher. With mediocre keyboarding skills, regular gazes that meet the center region of a common 15 inch laptop screen actually raises the eyes toward the front of the room. This equates to anything between a 10 and 20 degree reduction in the downward tilt of the head compared to students equipped with conventional paper notebook and pen—or worse, small cell phones with even smaller screens.

For the laptop users the more level angle increases direct and peripheral visual awareness of goings on in the teacher's direction and so potentially reinforces message redundancy by combining the verbal and non-verbal channels. This human-machine configuration also tends to allow the instructor to see the student's face and determine at least a baseline level of alertness (i.e., not drowsing behind a hand or hat brim).

However, there is a nasty downside to this equation. Near-vertical liquid crystal displays flashing image and light toward the back of a classroom simultaneously opens up the possibility for wider attention problems if other students are able to see (or are trying to see) that screen two rows up and to the left. To be sure, there is something quite moth-like about young peoples' relationship to flickering screens of all types these days. Even from more obtuse angles I have noticed a certain enticement in the faces of students besides and behind laptop users. Rather than attend to my instruction, they can at times be very intent on deciphering the screen activity of their classmates. So while having some positive force, laptops in class present us with potential attention problems in a few different ways. Of course, one easy fix for this would be to require laptop users to sit in the back rows. However, this is not an altogether satisfactory solution because it opens the door that much further for unrelated, off-topic, and/or inappropriate uses by those students with special devices.

It is not surprising, therefore, that a rising backlash against computers in class has been reported recently in *The Wall Street Journal, The New York Times* and other mainstream news outlets both on and offline. But should wireless machines be prohibited in all college classrooms? In specific kinds of classes? On a Wi-Fi'd campus, can this kind of thing even be efficiently and effectively proscribed? It is not at all clear that we can put the brakes on such activity, unless we prohibit computer use in classrooms altogether. But will it remain legal to do so? Technological innovation has provided our students with a new and potentially very powerful learning tool or, as the case may be, a very powerful distraction device.

Student attention to, and effective comprehension and recall of, things going on in the classroom probably did not suddenly dip in any appreciable way when the first laptops began showing up regularly in college classrooms fifteen or twenty years ago. Nor did attention necessarily plummet some ten or fifteen years ago when the early adopters of cell phones began toting their devices. It's likely, were it 1990, that this chapter would have been called "Doodling, Daydreaming, and the Discman." With some fairly subtle differences, the technical details of our digital distractions have stayed fairly consistent for almost three decades. The big difference today is that laptop computers, cellphones, PDAs, MP3 devices, and a growing cadre of portable personal digital devices have also expanded the possibilities for wandering minds.

As the saying goes, we live in an age of "ubiquitous computing." For so many people in the developed world, and now developing regions too, our daily interactions with fixed and/or small portable devices are becoming a seamless, perfunctory part or life. For example, we hit the saturation point of cell phone distribution a couple of years ago in the United States. Like the tobacco companies did decades ago, and automakers have been doing over the past decade or so, the leading handset makers and service providers are now altering their business models to meet and to make demand in these emerging markets.

While the same pattern is predictable with other digital technologies, we are not yet near saturation with portable computers and Wi-Fi networks here in the States. We are, however, certainly embedding these technologies throughout the country at a racer's pace. In some venues we are nearing saturation. Several years back a survey reported that 98 percent of the class of 2007 opted to purchase portable computers instead of desktops (Lyons, 2004). Today the last few percentage points are filling in fast, with wireless hubs and routers being enthusiastically installed in colleges and universities across the country and around the world. The theory is that laptop and tablet computers allow for a potentially more exciting, and certainly more interactive ways to enhance and enable the lesson of the day. The emerging picture, however, is that these tools may help us avoid and ignore as much as attend. Is it a zero-sum game? Time will tell. Though I have some inklings.

WHAT I SAW

In the spring of 2007 I sat in on a few popular courses at my previous school, a medium-sized private college in downtown Boston devoted to communication and the arts. I visited three small Fundamentals of Speech classes and one larger lecture section of a Media Production course. Positioning myself in the rear corners of the rooms in an effort to gain the best perspective of goings-on in real time, I paid especially close attention to students with Wi-Fi-enabled computers. The speech classes ranged between 17 and 21 students, whereas the media course had about 65 showing up on a typical day. In total, I counted 16 students using laptops out of the approximately 125 that came to all three classes. Of the 16 users, twelve agreed to participate in the study (2 freshman, 8 sophomores, and 2 juniors). Most were white, of European descent and just two were women.

Granting the introduction of some bias springing from these quasi-experimental settings (I helped a subset of the student participants install *Camtasia*, a screen capture program, on their computers), I was nonetheless surprised by their willingness to be observed and get involved in the research to the degree that they did. Of the twelve participating at various levels during my observations, five expressed overt enthusiasm about helping out with the project. These young men were especially keen on taking up an active role—seeming a bit proud, perhaps, to be cutting-edge, pioneering twenty-first-century citizens and students. This, however, may be an artifact of that college's reputation as a hip, high-tech kind of place. Or maybe students are simply growing accustomed to being surveilled in so many subtle ways throughout their day that they just don't care who's watching anymore. As one of my media students later put it: "Hey, you can stick a camera up my ass for all I care. Just give me that six-figure salary and go ahead and snoop all you want. I'm not doing anything wrong anyway." I think there are some problems with such an outlook that my young friend did not think through at the time, but I won't attempt to address those here. Instead, let's continue with what I noted while observing student activity in those classrooms.

Only one of the twelve individuals observed had just one "window" open (a word processor for taking notes) on their screen. All of the other participants ranged from maintaining three to eleven windows open at one time. Uses ranged widely, including activity on various e-mail accounts, Web browsers, campus library searches, WebCT discussion list activity and PowerPoint viewing, LAN-and Internet-based social networking, instant messaging applications, on- and off-line computer gaming, music and video downloading, shopping and browsing at various online retailers, *eBay* and *Craigslist* searches and transactions, as well as online banking and dating services.

As a representative example of the behavior and practices in this group I'll discuss details gleaned after observing an individual who, for one hour during an afternoon meeting of Speech Fundamentals, maintained a broadband wireless connection to the Internet. The topics of discussion in class that day were "persuasion" and "information and credibility" in reference to an upcoming informative speech assignment. This male sophomore audio major had nine tabs open for seven applications across the bottom of his screen. He kept five of these layered in a cascade of frames to better allow monitoring of activity in each:

1: a private e-mail account at Yahoo (showing a message from friend at same school)
2: a school e-mail account (currently showing assignment notice from instructor)
3: a Word doc file in the works as the note page for current class discussion
4: an AIM channel w/running commentary (with another friend at same school)
5: a Google search (looking up facts about RFID applications being discussed as potential debate topic by instructor)

The remaining, less crucial, because apparently inactive tabs included:

6: another Google search (that found a political satire site mentioned by another student, but also related, if somewhat tangentially, to ongoing class discussion about credibility).

7: ad w/pictures of mountain bike student was thinking about buying on Craigslist

8: a partially written e-mail message to a HS friend attending a college in CT.

9: a second Word document (rough outline of informative speech in the works).

OUR ATTENTION ECONOMY AND THE MULTI-FACETED NATURE OF MULTITASKING

What stood out for me in the midst of this student's deft handling of an ostensibly confusing array of information on the computer screen was that moment the instructor queried the class about the current state of affairs with RFIDs (a potential topic for upcoming debates in the class). In a matter of seconds the student seemed to move effortlessly from the midst of a casual AIM reply, to the inactive Google dialogue box, and came up with some useful details about subcutaneous RFID implants in pets, individuals under house arrest, and even some *speedo*-clad beach-goers in Brazil.

At first pass, this ability does not appear to be all that uncommon today. Some commentators are now even arguing that effective, efficient multitasking is part of the natural evolution of human consciousness. If so, I happen to be pretty bad at this kind of thing and worry about my role in the future of the species. Then again, my clumsiness with new technologies is likely due to my being, well, "old" by today's standards (I'm well into my forties). Truth be told, it also takes me a while to get up to speed with cognitive tasks like writing, and reading. Furthermore, once there (and when the getting is good) I tend to quite effectively screen off extraneous, ambient activity and sounds—often without intending to, and often to my own detriment. My wife, for instance, has been annoyed more than once by my professional ability to "fade out" while reading a book or article. Granted, one reason this is irksome for her probably stems from my periodic decision to bring a copy of *The Economist* or *Wired* or *European Car* to the dinner table with me. My etiquette, in other words, is certainly questionable in such instances. So here we have what appear to be two very extreme examples of multitasking; one that seems to work, the other which does not. But, we might ask, what is multitasking really? Like they say about pornography, do we *know it when we see it*? Concerning multitasking, is there a difference between a conceptual definition and an ostensive one? I think there is, and this means we likely have a lot of people who misconstrue what the thing entails.

Not long after collecting my initial classroom dataset, I came across a comment made by an up-and-coming efficiency guru named David Allen regarding multitasking that really struck me. Allen suggested that we are not experiencing a problem of information overload these days. Instead, according to Allen, most people find themselves at a loss in relation to the relevance and meaning of the information they have at their disposal. In other words, it is a problem of integration, of interpretation, and understanding. Organized properly, claims Allen, just about any amount of information can be brought into focus and under control. In one graphic, Allen's GTD (for "getting things done") recipe closely resembles a computational flow chart from the one computing course I took as an undergraduate in the mid-1980s. And as I'll suggest later, there is perhaps something telling about a computational model, and therefore a computational analogy, being marshaled to represent human thought processes.

In contrast to Allen's take, I found a characteristically succinct and pithy little quote from McLuhan's *New Media as Political Forms* (1955). McLuhan wrote: "it is the sheer quantity of information which has alienated us from political and social reality." Once again we seem to have two basic arguments in the offing. More recently we have Allen suggesting in his seminars that the human brain is almost infinitely malleable. And this notion of a highly "plastic" brain has certainly gained a great deal of traction in recent years with research centering upon the ability of aphasics and Alzheimer's sufferers to re-orient, re-tool, or re-train different regions of the brain. There is also a sobering caveat embedded in McLuhan's words, however. It is the idea that we are not infinitely capable, that we are still much more human than machine, and that there might even be something special and sacred about our *in*ability to attend to multiple things at the same time.

And so here again I want to point to a paradigm shift long underway regarding the way we tend to think about thinking. It began in the corners of certain academic circles. The Macy Conferences, for example, which continued for almost a decade beginning in 1946, were intended to develop or at least set the terms for a "general science of the workings of the human mind." The impetus for these conferences finds origin in questions raised by Norbert Weiner, Jon Von Neumann, and other pioneers of cybernetics, computer science, and information technology. One of the primary "sticking points" eventually brought input from philosophers, anthropologists, and early communication scholars like Gregory Bateson. It was, for some, just a nagging suspicion, for others a growing awareness, that the *human mind* is not simply the brain in the head. Rather, mind is an extended cognitive apparatus that includes the entire body, and a host of artifacts and accompaniments external to the body. Boy, if the Macy attendees could only see us now.

Certainly our current realities suggest something very different from what we believed in the past about what it means to be human. Gerry Moira, chairman of RSCG, a leading advertising firm in London, points out that multi-screens engage us at every turn these days, and that this is the future of advertising (Moira, 2008). I want to suggest that the theory has largely formed the reality. An economically prompted self-fulfilling prophecy has a lot to do with this. It is becoming increasingly clear that we do indeed live in an "attention economy" as Herbert Simon opined almost forty years ago:

> In an information-rich world, the wealth of information means a dearth of something else: a scarcity of whatever it is that information consumes. What information consumes is rather obvious: it consumes the attention of its recipients. Hence a wealth of information creates a poverty of attention and a need to allocate that attention efficiently among the overabundance of information sources that might consume it (Simon, 1971; p. 40–41).

Somewhat less distanced from it all than Simon perhaps, Moira admits this is a dystopic world in the making. We are slowly discovering, it would seem, that the all-at-onceness our multi-screens afford us is now coalescing into something potentially anti-human. More sinister and un-therapeutic than Simon's prognostication, it is a commercially based attention economy that is emerging, and one designed with a fatal flaw. One serious problem with our current media environment seems to be that it is not actually designed with such noble aims as "integration" and "understanding" in mind. And the economic dictate looms even larger in cyberspace than it does in the offline world of hard copy, billboards, and beamed and broadcasted sound bites. If I had to characterize the ethos of our new media world, it would be one geared toward dis-integration more than anything else. Consider, as one example, the modern notion of *news* discussed in the previous two chapters of this book. The nature of news as a competitive undertaking designed to fill space and time with content went fully into overdrive with the 24/7 "news cycle" that is generally agreed upon to have been conceived along with the birth of CNN. I placed the phrase "news cycle" in quotes in this case because it is no longer really a cycle at all, for there is no *regroup* or *down time,* per se, that would allow for any recap or reconsideration or heuristic reflexivity. Efforts between content providers to out-scoop, to out-update, or to find some other distinction that sets one source apart from the other now reign supreme. And this certainly contributes to our global attention deficit. Recall that media professor's comments in an earlier chapter. She suggested that her students aren't paying attention to the right things: "we're raising—and graduating—students who are totally unengaged with the world beyond the latest sports and entertainment headlines."

However, the lack of engagement, or the inability to engage, that this teacher describes may not be what it appears. To be sure, students are attending to things in the news, it's just that there is often too much of it that is disparate, and which is thrown at them without much effort to integrate or show connection. The language Neil Postman once used to describe TV news, works just as well, perhaps better now, as a description of our increasingly visual-oriented Internet-based news and information. In short, what seems to be most readily available in our information ecology is a streaming series of *ands*, with precious few *becauses*. A sense of connection, of relation, and even causality is hard to come by if we limit ourselves to inputs from video screens.

But these attention problems associated with our multi-screens not only present us with new ways to miss important information. I refer here to the way it can sometimes be difficult to avoid certain kinds of information too. Depending on the issue or topic or event, some things can, for a spell at least, blanket the media horizon. Such instances represent new forms of anxiety that far outstrip the confines of the college classroom, and can function in an almost tyrannical way as one respondent suggests. JJ, a 38-year-old professional from Boston, recounted the angst and excitement she and a close friend experienced in the lead-up to the opening night of a film they were planning to attend in high fashion. The two women were nervous about accidentally being exposed to details prior to seeing the first *Sex and the City* movie. As she put it: "Video clips, sound bites, and written snippets always seemed to pop up at the worst times. In the corners of TV screens, in text messages from friends, even in a sort of sinister way in the subject lines of some e-mails. I swear some people were trying to spoil the plot for me. We had to stay away from the information . . . I mean actively avoid it." This is a pretty poignant take on modern times. With such experiences the attention economy indeed becomes almost tyrannical.

I had my own interesting experience, this one on the "high seas," that further illustrates a kind of technological tyranny at work that affected both the primary technology user I was monitoring, as well as his onlookers. It also extends JJ's insights regarding attention and the emerging anxiety issue. On a sailboating excursion out to the Boston Harbor Islands I had the opportunity to spend more than seven hours with a well ensconced *Blackberry* user. George, a 32-year-old architect, brought his pocket-sized digital assistant along with him to keep track, most notably, of any potential messages from a young woman he had recently met during a pub crawl.

For the first couple of hours things went more or less as expected on a 27-foot boat carrying four thirty-somethings, two coolers filled with beer, and some grillables. While I sometimes felt alienated by all the cable TV talk (with a solid hour devoted to shows like the *Simpsons, King of Queens*, and *The Wire*), I readily engaged my shipmates when news, politics and movies were the topics of discussion on deck.

About three hours into our voyage puttering around the shallows, I noticed that George began to flit in and out of phase with the rest of us, sporadically checking the status of his e-mail and text messages. He came in and went out of the conversation, sometimes smoothly, but more often off-cue, and almost ADD in effect with multiple media references: the antics of Homer Simpson, the main character Doug in *King of Queens*, and a couple other characters from prime time shows within a short time span. George's behavior had the look and feel of the antics and rantings of a hyperkinetic six-year-old—and I know a few of those too.

A good conversation flows both for the participants involved and from the vantage of those at the periphery. It's about being in-context both in mind and in body. This integration or perceptual binding of the agent in time and space is a large part of what enables genuine dialogue to occur. And here again, we know it when we see it. It embodies the *I/Thou* portion of Martin Buber's *I/Thou-I/It* distinction, and now communication researchers, phenomenologists, and cognitive scientists are finally converging on the idea that a rigid distinction between mind and body only muddies our efforts to understand human cognition. That distinction happens to be an unfortunate legacy of Rene Descartes. Cartesian dualism ends up being a debilitating doctrine, and mode of thought that artificially separates out the mind (or, in this case, brain) from the body. I've argued throughout this book that we are much more holistic and interconnected than Descartes was ever willing to believe. While he disagreed vehemently in a sort of follow-up/debriefing interview, George made it abundantly clear in-context that he was having trouble switching between his two communicative environments.

HUMAN CONTEXT SWITCHING, MULTITASKING, AND THE MYTH OF PRODUCTIVITY

The term *context switching* was borrowed directly from computer science and is a fundamental characteristic of any digital computational device. Relative to other processing modes, context-switching is computationally intensive and hinges on maximizing the quantitative efficiencies of recursive, electronically enabled information processing. In layman's terms, context-switching allows multitasking, which is one way to make efficient use of a limited amount of time. One misapprehension about computers, however, is that their CPUs are continuously engaged in multitasking. Another is that multitasking means or equates to "simultaneous processing." It does not. The fact of the matter is that any single-processor system functions task-specifically (and this is one reason why we have seen dual- or multi-core processors hitting the market lately). But given that integrated circuits process near the speed of light, our impression is that they are always doing many things at the same time. Again, they do not. *Interrupt handlers, time slices,* and *switch frames* are still part of the standard nomenclature referring to the underlying computer architecture that betrays the reality of these machines running discrete processes in sequence. Again, from a human's vantage point, it all appears to be happening at the same time, but this is merely an artifact of how beings like us perceive the world.

Champions of human multitasking like David Allen often cite driving, or flying planes or helicopters as proof positive of our wonderful ability to multitask. The problem with these examples is that they all describe what are in reality single tasks, or at least multiple sub-tasks with a single end or overarching task in mind. And therein we discover the key to understanding just what multitasking entails. Driving a car, for instance, is really a collection of smaller sub-tasks, though the subtasks are becoming fewer as automated systems take over some of that cognitive load (like shifting gears, breaking, manipulating lights and signaling controls, and even parallel parking). But from the *Model T* to the latest gas-electric hybrid, once we know how to control the car, driving is no more multitasking than walking.

The definition of multitasking, then, hinges on two things: the nature of the task and the level of conscious attention required to perform the task—or to successfully switch between sub-tasks as is the case with driving. Driving an older car, say, one without ABS but with a manual transmission (the most complex kind of driving one can do these days) includes clutching, braking, shifting, accelerating, looking down the road, in mirrors, over shoulders, etc. But when we look at many typical drivers on the road today (despite the mass exodus toward automatic transmissions and systems) we find that the tasks now associated with "driving" are actually much more disparate and discrete actions, like having a conversation with a passenger, adjusting the radio, talking on the phone or texting, finding a track on the iPod, etc. All of the above and so much more can inhibit the flow of that experience that is driving—the embedded, embodied experience of almost being at one with the vehicle. And this is when the trouble starts. Even the act of listening to a news item on the radio, or trying to track a fellow passenger's recounting of some event has been the trigger for collisions big and small. There is a phenomenon which accounts for the widespread inability in the population to perceive how substantially the physical and symbolic environments affect us. Mass media researchers call it the third-person effect.

The third-person effect refers to the awareness, by many, that some factor or force in the environment (like a medium or a message) can be incredibly powerful, and that most people are affected in some way or another by it. The counterintuitive part of the story is that most people are pretty sure that *they* are not the ones being affected. I think the third-person effect probably looms largest with those who regularly attempt to multitask while driving. What we actually find, however, is that drivers who think they are doing a great job are often travelling 10 to15 mph under the speed limit or are simply careening all over the road. In the context of these new media forms, what we are discovering is that media users who report being efficient and effective multitaskers in reality end up being surprisingly bad at what it is they think they are doing.

A college senior and news director for a popular college radio station centered in downtown Boston certainly falls into the third-person effect group. While he thinks he is among the fifty percent (or so) of the population who believes they can multitask productively and efficiently, his remarks boldly betray him, especially given his position as a budding news professional: "I'm not trying to get it all" he says "just the gist." There is a sad irony here. Should a news production student just be interested in getting the gist? Another senior at a different school in town is editor-in-chief of the award-winning student paper. She is positively certain that she has learned to multitask productively.

First of all, (and I gleaned this from their own descriptions) these young people do not seem to be multitasking during most of their time on the job. Certainly not in the way an iPod-manipulating driver would be, or George on the boat was attempting to be. The radio director describes trying to pull in various leads in the process of making a news report. In other words, all have a single task, process, or end in mind. Likewise, the paper editor describes editing a document while fielding text, instant-, and e-mail messages related to that document. Nor is this multitasking per se. Why? Because neither of these individuals are actually switching contexts (in this case, cognitive and experiential contexts) to any substantial degree.

Or, as Clifford Nass puts it, "learning how to do single things more quickly, but still in series" is not multitasking. Nass and his team at Stanford recently completed the first phase of a multi-phase research project that looks into the underlying nature of multitasking. And the context-switch seems to be the litmus test. Once again it is George who is multitasking (or is trying to) when he's flitting between a conversation with the guys on the boat and texting his female friend about a hopeful date. And it is the student attending a lecture while checking the latest sports scores on his iPhone, and clearly the driver speeding down the road while searching for her favorite driving playlist. These are bona fide multitaskers, however inefficient, ineffective, and/or dangerous they may be in the midst of the various tasks under way.

So there is clearly widespread misperception with respect to what multi-tasking entails—and when someone is, or is not, in fact multitasking. Coupled with the third-person effect, this misapprehension creates a very confusing situation indeed. While there is often bona fide multitasking happening in fits and starts all around us, Nass remains doubtful about our ability to get better at it. As his research to date shows, there is virtually no evidence that people can learn to handle discrete multiple tasks simultaneously (Ophir, Nass and Wagner; 2009). To the contrary, multitaskers seem to only get worse over time. And the literal disconnect is the context switch. Context switching of the sort enacted by George on the boat, or by my students on their computers and cell phones is (while not physically dangerous in the way texting while driving can be) nonetheless disruptive of the experience of cognitive and mnemonic flow.

SOME THOUGHTS ON THINKING

I've taken a lot of candid shots featuring my six-year-old son drawing, working through a puzzle, manipulating clay, and building models with the tight focus of a yoga master. I love those pictures. I'll hold onto them because there is something incredibly serene about these moments that I also want him to remember. But I also want to remember that feeling of oneness, of being-in-the-moment, as such moments seem to be increasingly rare and fleeting these days. There's no question, I often feel as though I'm losing my ability to sustain that kind of attention in my own work and leisure activities. Preoccupations, ongoing events in the periphery, mental fixations and day-dreaming surely distracted and debilitated our ancestors in various ways at least as long as we've sported bi-cameral brains. But something new is afoot.

Since I left for college in 1985, and until 2009 in fact, I never had a television set hooked up to any cable or satellite network at home. Nor did I ever have a real Internet connection. I wrote toward the beginning of this book about my early-90s experience e-mailing into my University's mainframe, but I only maintained that text-based connection to the Internet through the end of my Master's program in 1994. After that, my 300-baud modem was perceived to be sluggish enough to warrant my walking over to the campus a few blocks away and log onto one of the fast terminals in the library or a computer lab. I've experienced fast and generally reliable high-band access to the Net at my school office for just over a decade at the time of this writing, but I never had a "domestic" graphical interface connection until mid-2009 either. Of course, until they locked it recently, I could sit on the back porch and periodically catch one of my neighbor's wireless signals with my laptop, and my wife's iPhone always enabled a quick e-mail or basic Web inquiry in a pinch, but I have been more or less "off the grid" with respect to the domestic context in comparison to most people I know for the last decade, even two. And while I have tried to limit my son's exposure to the various devices that are either designed to, or accidentally help us, "multitask" these days (or just provide the opportunity for diverted attentions), I know that the information environment enabled by these new technologies will eventually seep into the periphery of his awareness. Whether it's the onslaught of endless information, or the digital appendages he will undoubtedly become so familiar with, these artifacts will find their spaces in the corners of his mind.

And so an important question needs to be answered—and I think with all deliberate speed: What is it to be deep in thought, and what value does this kind of behavior, this way of being, hold for us? Like my son engaged in his artistic endeavors, what is an engrossed reader, an ensconced listener, or a rapt watcher really doing? What's more, how might the resulting thought processes and patterns differ from the experiences of the iPhone'd, Wi-Fi'd, texting and twittering generations coming of age? Again, I know my own son will eventually take part in much of that and so will become familiar with these new ways of thinking, communicating, and being in the world. I want, therefore, to do whatever I can to enable him to periodically get back to that earlier mode of life, a mode he still knows so well.

There is a long history to such questions of mind and body. Informing modern flow studies is the *mindfulness* of Zen-Buddhist and Taoist teachings and the *mindlessness* described in the Bhagavad Gita and Upanishads. In the West we have the writings of Heidegger, Merleau-Ponty, and James. The basic idea behind it all is doing one thing at a time, and doing it well, before moving on to the next thing. It is about *being there* according to Heidegger, or being mindful of *here, now,* and *this* as the Taoists liked to say. The point is that attention and focus have been of special interest to humans for a very long time. William James, in his *Principles of Psychology, Vol.1* (1890), puts it well enough:

> Everyone knows what attention is. It is the taking possession by the mind, in clear and vivid form, of one out of what seem several simultaneously possible objects or trains of thought. Focalization and concentration of consciousness are of its essence. It implies withdrawal from some things in order to deal effectively with others, and is a condition which has a real opposite in the confused, dazed, scatterbrained state which in French is called distraction, and Zerstreutheit in German (p. 403).

Even if James idealizes a sort of perfect and full attention in that quote, most of us nonetheless know what engaging in full attention is like. And we still might even long for this sort of thing. To be sure, images depicting entranced, highly focused individuals lost in thought are staples in just about every culture on Earth. Consider a rendering of the *Buddha*, or Rodin's *Thinker*. The expressions are reminiscent of a seasoned reader deep in the pages of a good book. But they also remind us of a rapt television viewer, or attentive video game player hooked into their machine. Indeed, these images seem to do just as well if single-mindedness is what we're after.

However, it is not unusual if one of two things is occurring when someone is engaged in attention states of the sort just mentioned. In some cases the person is focused in such a way that the flow of mental and physical focus is in response to a cascade of external stimuli impinging on the emotional centers and limbic regions of the brain. This is most often the case with a non-discriminating television viewer—the most common kind of viewer in America according to George Gerbner's "cultural indicators" database. At other times, say, during deep reading, the logic center is at the helm, leading cognition with a robust series of interactions between the left and right hemispheres. And now the philosophical and neurological significance of this is coming to light, with functional magnetic resonance imaging (FMRI) showing us where, when, and maybe even how the brain does its work. But then again, all the high-tech is not always necessary since, when the brain is working well, we sort of know it when we feel it. Or, if one needs to be reminded, consider some of the key elements Mihaly Csíkszentmihályi identified as being characteristic of *flow*:

- *Concentrating and focusing*, a high degree of concentration on a limited field of attention (a person engaged in the activity will have the opportunity to focus and to delve deeply into it).
- *Direct and immediate feedback* (successes and failures in the course of the activity are apparent, so that behavior can be adjusted as needed).
- *A sense of personal control* over the situation or activity.
- *The activity is intrinsically rewarding*, so there is an effortlessness of action.
- *Individual becomes absorbed in their activity*, and focus of awareness is narrowed down to the activity itself (*action + awareness merges*). (Csíkszentmihályi, 1990; 1975. p.72)

To extend this, we could even distinguish *flow* into both internal and external components. Internal flow would describe the internal or subjective or phenomenological aspects of feeling in tune and in-sync in the midst of a particular task (whether that would be reading a book, watching a movie, attending a college class, or having a conversation). External flow, in turn, would be the more physical experience of moving through space and time in a smooth and deliberate way (and perhaps the perception and feeling by others too that you are doing so in their presence).

With a solid understanding of his Tetrads, or laws of media, I've come to appreciate McLuhan's "predictions" regarding media use. Unfortunately, his brand of technological determinism has been badly distorted over time, with nearly four decades of restatement occluding much of what he was trying to say about the causal role of communication media in particular. Toward the end of his life McLuhan made considerable effort (with the help of his son Eric) to answer critics' charges that he was probing aimlessly through a thick fog, coining hapless aphorisms along the way with little intelligibility, and even less predictive power. But I happen to think that McLuhan's ecological approach to media studies is much more useful when framed as a prescriptive theory, rather than the predictive one he was hoping it might be until his death in 1980.

Before drawing out the prescriptive nature of McLuhan's work, I want to first remind the reader that McLuhan did foresee quite clearly what I am writing about here. In the late 1960s with *The Medium is the Massage,* and *War and Peace in the Global Village*, he predicted a kind of "global ESP" emerging as an inevitable consequence or by-product of computer technology. McLuhan was essentially implying that a media system would be created that is very much like what we now see in the mobile Wi-Fi computer connections that are springing up all across our campuses, cities, and neighborhoods.

There is a profound sense too, and it's just sort of hanging in the air these days, that this kind of technology is increasingly necessary to both work and life. A recent *Sprint HTC Evo* television commercial says it all. In a thirty second spot we're offered one take on the history of technology. The ad is remarkable not only for its impressive condensation of that history, but also in its hyperbole regarding the cultural status of the now ubiquitous smart phone (and perhaps Sprint's in particular). In a dominos-like cascade of technological artifacts beginning with a stone wheel, we see locomotive trains, a phonograph, a typewriter, telephone, automobile, televisions, airplanes, and Saturn V rockets (among many other iconic tools) toppling toward the future. This story of innovation culminates, of course, in the little Evo 4G, with its solid 5-bar connection still standing at the end of the line.

However tongue-in-cheek the writers of that advertisement may have been feeling at the time, I believe they have tapped the right zeitgeist. Indeed, it seems safe to say that a substantial portion of the population is drunk on this dream of modernity and its *newer = better* edict, with digital communication appliances representing the sine-qua-non of human achievement. The smart-phone becomes the talisman of our collective dreams and aspirations. In his most positive moods McLuhan also appeared hopeful that a world-wide communication network, perhaps accessed through such devices, could create a new level of empathy around the world. But more cynical undertones are found elsewhere, and throughout most of his corpus he seems doubtful regarding any real liberatory potential in new media.

My impression is that McLuhan did, and would still, think it's pointless to try and halt technological innovation. And I tend to agree. Indeed, it seems unlikely that we can stop something like the proliferation of cell and Wi-Fi networks and systems. Regarding such hopes, I always like to point to Neil Postman's comments about television. In the epilogue to *Amusing Ourselves to Death* (1985), Postman likens any efforts at "pulling the plug" on TV to yanking the skeleton out of the human body. Unfortunately, in both cases, everything would fall apart were such schemes even possible. With respect to cell and Wi-Fi systems, the results could be still more dire at this point.

Of course, it's not all the fault of the attention deficit of citizens, students, employees, and friends. A constant prompt that helps explain how we have come to this quasi-symbiotic embedding and interweaving of humans and their technological artifacts, is that the telecoms have a long history of prognosticating with precision both where and when a new system will catch fire. There is a certain truth to the *"if they build it we will sign up and log on"* mantra of late. Of course, peer groups and other social pressures have something to do with it, but powerful network effects certainly come into play too. And this is perhaps the ugly side of McLuhan's technological determinism that is so simple to see—and, unfortunately, so easy to gloss. It is the part of his thinking that used to get critics all riled up. But few would now argue that this is where we presently find ourselves—in the midst of an attention-getting, attention-turning, always-on mode of life.

Nass and his Stanford team have noted that more and more workplaces are being designed specifically to foster the habit of "multitasking." In the worst-case scenarios, as Nass would have it, employees are being required to keep multiple chat windows open. But this is not, ostensibly, to talk to their friends. The theory is that if the lines of communication are left open between co-workers, then both the quantity and quality of work will benefit. The question Nass is asking is an ecological question. He wonders whether we are structuring work, and the workplace, in ways that are impacting negatively upon our minds and brains (Ashbrook, 2009). More than an ecological puzzle, though, what we have here is a distinctly media-ecological question. The McLuhan meme is finally going mainstream.

Nass points out that when you ask the brain to do more than it can do it doesn't always choose the best route for each task. And so more often than not we end up doing everything under par. And as for the suggestion that we can get better at multitasking, Nass offers the analogy of a car. When you push an automobile too hard, it doesn't learn to run better. In fact, everything tends to get worse. If pushed to its outer limits the engine can overheat, pre-ignite, and become both erratic and unstable. And certainly where high speed or extended driving are involved, a host of associated systems can begin to fail. Transmissions can blow apart, brakes overheat and begin to fade, tires loose traction or rupture.

I like Nass' comparison to the car because, again, while there are limits to the analogy, the brain can also be conceived as a sort of a machine. At least up to a point, it can be very useful to think about thinking in this way. Human beings with perceptual and cognitive apparatuses spanning brain, body, and world are ubiquitous today. Just think about the blue-toothed pedestrian making his or her way through the streets: walking and talking, drinking and thinking and driving. And when we consider the many ways we are now hooked in to machines of the digital sort, the capacity that collective (though still at bottom biological) apparatus has to keep up with the onslaught of data, information, and other stimuli is understandably pushed to extremes. In these cases, however, our biology is the bottle neck.

Marcel Just of Carnegie Mellon's Center for Cognitive Brain Imaging thinks multitasking may be the "coal mine" of the modern workforce (ibid). Dr. Just points out that some things just can't be willfully kept out of consciousness, "if someone is speaking and you can hear it, it gets into your brain and consumes resources." To be sure, this is just the way language using creatures like us respond to code-specific linguistic stimuli. Just qualifies this somewhat by suggesting that it is possible to hear two simultaneous streams of speech and follow what both are saying, but generally, once we take on two tasks, we don't do either as well as if we went with one alone (ibid). The problem, says Just, is that "we get brown outs . . . there are limits. As you try harder to increase the number of tasks you process, you end up allocating fewer resources to each of the participating tasks" (ibid).

So an unfortunate irony seems to be built into our bones: human beings today are, more than ever, associative engines. And while there are certainly great benefits to being adept at making all kinds of connections between the stimuli impinging on our senses, there is an important and often over-looked downside as well. The biggest problem with free and constant association (and the new tools which enable this) is that it makes us into really bad filters for irrelevance. In a previous chapter I detailed the manner in which the latest search engine technology seeks to bolster this shortcoming only to result in a new debilitation—information overload.

THE HUMAN AGENT: ALWAYS A QUESTIONABLE MODEL OF AUTONOMY

Where, then, is human agency in this new media equation? It is a difficult question, but perhaps not as impenetrable as the question of how to subdue the corporate, top-down, network Goliaths that threaten us. No, we can't stop the factories and assembly lines. Intelligent machines are now making even smarter machines—and faster than ever before. This is an essential piece of our Digination too. And to wish for an end to it all is just wasted will. It is possible, however, to buck the system by learning to make strategic choices. I am talking about media choices: the right tool for the task in the right place and time. And what is most hopeful in all of this is that the prescription might end up being a fairly simple set of rules.

One recipe for action could be that vaguely eastern approach to life that also has some common sense roots: The notion of getting things done—of attending to and completing the task, however short or small. It's about attending to *"this, here, now"* as Taoist teachings suggest, and can be so much more than pie-in-the-sky ponderings. To this end some things are certainly getting accomplished to curb our mythic multitasking propensities. Since it is a kind of environmental issue, the idea is that we can and need to start thinking about different ways of structuring our work, and our workplaces.[2] The bet is that the structure of our minds will, over time, follow suit, or so new work in neuroanatomy and epigenetics seems to suggest.

From neuroanatomy we are learning about cell structures in the pre-frontal region of the neocortex called mirror neurons. It is the part these cells presumably play in representing cognitive, emotional, and physiological states, and triggering such responses, that has researchers in human-computer interaction, cognitive science, robotics, AI, and Media Ecology paying close attention. Coupled with epigenetics, or the idea that our genetic makeup can change due to what comes in addition to or after our original DNA sequences are laid down, we see a very intriguing story taking shape. Epigenetics had a shaky start near the end of the twentieth century as it was seen as a kind of return to that Lamarkian heresy of intra-generational change. Indeed, it seemed to rail directly against the orthodoxy of Darwinian selection and virtually all conventional wisdom regarding speciation following from that. Today epigenetics is enjoying a vigorous renaissance, and this holds profound consequences for the kinds of natural, physical, and symbolic environments we are altering or building all around us and, by extension, who we are becoming as a result of these environmental changes.

All these changes notwithstanding, we are also discovering that there may be limits to what we can do with our brains, and so what we can become. This does not detract from the fact that our ancestors were slowly augmenting that brain since the first bone shard or stick was poked into a hole or jabbed into flesh, but it should prompt us to look at the kinds of tools we are merging with today. Certainly, if the push toward multitasking in all of its guises continues unabated, we will change the structure of our cognitive apparatus. That is, we will learn to adopt and adapt.

However a *Verizon Wireless* advertising campaign featuring the *Droid 2* seems to forget all of the nuances. It also ignores any sense of limits. The scene is a high-tech boardroom where a meeting is in session. A smartly dressed 20-something attendee is listening to the spiel at the end of the table. Our young professional then spies his new Droid phone on the table in front of him. Presumably, a pressing message has just arrived. He orients to some technical schematic, and begins typing. Over the next 10 seconds his hands and forearms morph into mechanical appendages. His now augmented thumbs tap out the message at a blistering pace. The narrator then reminds us that "this droid has evolved into even more . . . turning you into an instrument of efficiency." It's not clear to me, by the end of this 30-second spot, who's running the show here—the Droid or its human counterpart. Extending this idea, another Verizon spot for the new Droid contains an overdub that includes the phrase "it's hard to tell where you end and it begins." There is a delicious irony in that ad copy too. And indeed, as some of my Mohawk friends liked to say, *adaptation does not mean assimilation.* And perhaps it *should not* either. We have always lived in a social construct, and we are certainly in the midst of ongoing evolutionary adaptations, but the environments and niches we are now constructing are of a kind we have never dealt with before. For this reason we need to learn to choose which tools to merge with in this Digination.

I am stating the obvious when I say that the American workplace, its educational institutions, and so many of its public and private spaces have become saturated with personal digital media and communication technologies. But the beginnings of a cultural backlash and a sort of urban environmental movement may be underway. One of the interesting patterns now emerging is a collection of new prohibitions being put in place in many corners of corporate America. A few of these appear counterintuitive, like not using your phone when walking down the hall. Others are interpreted to be downright Draconian. It is not uncommon, in fact, to find baskets in the meeting areas of corporate offices, where iPhones and Blackberries must be deposited prior to entering the inner sanctum. But this new way of thinking about attention and the portable media use that can perturb it is also trickling out into other sectors of mainstream society.

A senior communication major who works at a popular water park on Cape Cod informed me of a company policy recently put in place at the facility. Any employee who is caught with a cell phone merely in his or her possession while "on the clock" is summarily shown the exit—fired with no discussion. To date, all have complied. The severity of the policy reflects the very serious state of affairs we now find ourselves in. There's no doubt about it, things like texting and twittering on the job have become serious safety issues, with inattentive automobile, bus, and train drivers making headlines for causing more than a few casualties. In less life-and-death scenarios too we are seeing new rules and regulations taking shape. But while blanket prohibitions are certainly one strategy for dealing with such problems, there are other ways to structure our media environments so that we get the best possible performance from our human-machine interactions and interfaces. Many of these new prescriptions revolve around much more subtle, context-dependent, and task-specific details.

There have been recent acknowledgments by a growing number of educators and researchers that computers in the classroom are much less effective or productive than initially hoped (Wurthheimer, 2003). We are learning, albeit slowly, that just because something like 24/7 access to the Internet is now possible does not mean that we ought to be maintaining such access at all times. The moral of the story? Technological progress does not always translate to social progress, and it certainly does not always imply epistemological progress. Knowledge and information are not the same things, nor is having a memory somewhere on board (i.e., the "possession" of information) the same thing as having wisdom or intelligence. Now, more than ever, we should be wary of thinking otherwise and taking the "numb stance of the technological idiot" as McLuhan liked to say.

But this is often easier said than done, for it requires some collaborative work. We can cajole them all we want, but the young will not relinquish their machines in response to brute dictate or even the most subtle rhetorical ploy. Logic and argument will do no better. They need to experience it for themselves. As teachers, parents, siblings, employers and friends, we need to help show them what it's like—how it feels to periodically unplug. To pick up that book, to slow down and think about something—anything—for more than a few seconds, to have a face-to-face conversation, to write something on a piece of paper, to send an e-mail longer than a sentence or two, even to use their smart phone as a simple phone more than just a few times a week, to be comfortable inhabiting both time and space, to feel it in the flesh, to be comfortable with themselves in the here and now. As a magnetic phrase I pieced together on my fridge reminds me when I grab the milk carton and break that rule of good decorum: "Wired media fiddle with the folk world." The prescription, while only implied, would read: "expose your kids to many modes." They need to feel it in their bones.

Of course, non-mediated existence is impossible. It always has been. To think otherwise is self-deluding. We must work with what we have. The young will only follow if we show them well that there is a good feeling, a *visceral value* in working through a conversation, in completing that paragraph, in finishing a task—no matter how small. To be in the moment in both space and time, they will know what *being there* means. But to get there, we have to begin again to associate positive things with "down time," with turning off and tuning out, with being quiet with oneself and others, and with taking it slow. To be sure, we have to teach our children well. We must instruct them in the art of attention, for in the midst of an environment teeming with digital distractions, there are still ways of getting back to just *being there.*

Like any tool, the Internet-connected computer, the PDA, and the smartphone have their own logic and grammar. But now there is a "convergence of kind," and so a convergence of telos or *end-in-mind.* Common to all of these new tools is an underlying nature that exerts a repeated demand on their users. For such devices, along with any human being connected to them, the *end* becomes that of amassing information, data, tidbits, and the new—with the implicit understanding (or is it an assumption, a hope?) that it will all be sorted out later. I think again of *Maslow's hammer* and put a contemporary spin on the phrase: to a person with a digital device everything looks like information to be downloaded, copied, stored, scanned, condensed, or otherwise processed and manipulated. And small, powerful, portable machines with extended battery life and almost limitless data capacity prompt us to want to process that information here and now . . . no matter where and when. These are the ways of being that characterize our Digination.

An awareness of the problems associated with our transparent relationship to so much digital media is illustrated by the recent introduction of several new attention and time saving applications available for a small fee (or for free) to anyone who is feeling the squeeze of their machines. I'll let the reader detect the delicious ironies contained in the promotional snippets for two such applications; *Freedom* and *Anti-Social*:

> Freedom is a simple productivity application that locks you away from the Internet on Mac or Windows computers for up to eight hours at a time. Freedom frees you from distractions, allowing you time to write, analyze, code, or create. At the end of your offline period, Freedom allows you back on the internet.

You can download Freedom immediately for 10 dollars through either PayPal or Google Checkout.

Freedom enforces freedom; you'll need to reboot if you want to get back online while Freedom's running. The hassle of rebooting means you're less likely to cheat, and you'll enjoy enhanced productivity. Freedom does one thing and it does it exceedingly well: It helps you get work done.

(emphases in the original, macfreedom.com/page/2/)

Anti-Social is a neat little productivity application for Macs that turns off the social parts of the Internet. When Anti-Social is running, you're locked away from hundreds of distracting social media sites, including Facebook, Twitter, and other sites you specify. With Anti-Social, you'll be amazed how much work you get done when you turn off your friends.

When Anti-Social is running, the only way to get around the block is by rebooting your computer. As you will feel a deep sense of shame for rebooting just to waste time on Twitter, you're unlikely to cheat. A registered version of Anti-Social costs $15.00, and a free trial is available.

(emphases in the original, anti-social.cc/)

These are no joke. So many people are having trouble disconnecting, and knowing where and when and with whom they should do so. As with a variety of top-down efforts in academe, at home, and at the office park, there is a growing recognition that modern digital media users might be having some trouble thinking, staying on task, and getting things done. If McLuhan did not say much about the approaching fog, there was some portent that goes at least as far back as McLuhan's heyday.

In drawing out his ecological approach to visual perception, the eminent perceptual psychologist J.J. Gibson (1966, 1979), suggested forty years ago that with all of the small screens (primarily television screens) then at our disposal, that we were well along the path of "living boxed in lives." If Gibson could only see us now with our Blackberries, Droids, and iPhones. If reserved for the upper echelon just a few years ago, these devices are fast becoming standard equipment—perhaps soon a natural right—for more than just the business executive today. Indeed, so many of us are now living in and through these small screens. It is as if we are reverting back to some pre-toddler phase of visual perception (or perhaps this all marks a return to the perceptual experience offered by something like Edison and Dickson's late nineteenth century *kinetoscope?*)

It is notable that by 3 or 4 months most human babies begin to crane their necks in an effort to follow whatever objects are moving to the periphery of their visual field. I refer here to something that occurs even before the capacity for object-permanence. Babies crane their necks due to a growing awareness that they have the capacity to expand their powers of perception by doing something intentional. They do this to figure out what's going on (where that face or object went, who or what is making that sound, etc.). They do this to learn new things about their world and solve questions that arise as they ambulate their way around the environment. So are we reverting to a synthesis of naïve and virtual realism (Heim, 2000) with our sticky screens? The new *WYSIWYGs*? Is what we see becoming all we seem to *get*?

NOTES

1. Consider the unveiling of the 4th generation iPhone in June of 2010. As Apple head Steve Jobs was getting ready to demonstrate some of the sexier features of the new device—in particular the video-call application—all hell broke loose. His wireless loop to the projection screen up at the front of the massive auditorium (and out across the ether to his design chief in England) froze solid. The auditorium filled with hundreds of Apple devotees and technophiles of various stripes began to stir. At first Jobs calmly remarked merely that "we have a problem." Then, in an uncharacteristically uncool manner for the mock-turtle necked, blue jeans and sneaker-clad billionaire, he began taking on the guise of a moody school marm, insisting, in a progressively frantic and condescending tone, that the audience needed to immediately shut down their laptops, netbooks, and indeed, their Mi-Fis, iPhones, and iPods too. Of course, at this point, the bulk of them were blogging, texting and twittering away in a massive feeding frenzy. There was some snickering, and plenty of murmurs reverberating through the crowd. Jobs, for his part, was not amused. Many folks present seemed to think he was off his rocker. His audience, most of course very friendly to all things Apple, had trouble comprehending the request. The irony was about as thick as molasses in January. The event illustrated something we don't think much about these days—how our digital devices have become transparent extensions of our senses and our selves. I now use a video snippet from Jobs' presentation in my fundamentals course to illustrate how one should never become overly dependent upon technological accompaniment in a public speech situation—indeed, Jobs did not know what to do—but there is also something almost nonsensical about such advice these days too. *What, no technological accompaniment?!*

2. Clifford Nass has recently become hip to this idea that was put forward by one of McLuhan's chief sources of inspiration—Patrick Geddes, the biologist-turned-urban planner who lived through another period of fast-paced technological change, the turn of the nineteenth and twentieth centuries.

Chapter 10

eBay Ethics

Prefiguring the Digital Democracy

"Human . . . is an adjective and its use as a noun is in itself regrettable"

—William Burroughs

I've addressed a lot of seemingly disparate facets of our emerging Digination so far. From a group of Mohawk's experiences with e-mail, to the underlying logics of popular search engines, to the changing natures of memory and attention our new digital appendages entail. In this chapter, we take a close look at *eBay*, the largest auction and commodity exchange venue ever conceived. I consider the core features of *eBay's* user interface and illustrate how, particularly in digital contexts, top-down (i.e., corporate, quasi-governmental, administrative) entities can very effectively enhance their abilities to rationalize and control what are ostensibly democratic, bottom-up, and peer-to-peer situations. Indeed, this penultimate chapter is not about *eBay* alone. It also highlights important relationships between the concepts of citizen, consumer, and socio-political actor today, and considers the significance these social roles might play in a full-fledged *digital democracy* of tomorrow.

eBay has set the template for so many online consumer-exchange hubs. In their efforts to construct a stable and secure digital marketplace, *eBay* designers may have inadvertently created a workable template for an efficient, if perfunctory neoliberal enclave. Doing well on *eBay* means much more than just getting a "good deal." It also means *doing good*. At one level, the business ethic on the Internet's largest auction site reduces to the appearance and feel of honest, timely, secure, efficient, well-packaged commodity exchange. However, with an increasingly sophisticated text-based feedback/rating mechanism at its core, successful interaction on *eBay* also requires that one be a good neighbor—that is, quick, honest, courteous, and kind in a way that might even be described as hyper-consistent. It is arguable that full-fledged and ideal citizenship on *eBay* is not wholly attained unless one becomes a "power" seller or buyer with comparable social capital. Both require maintaining high feedback scores. With all the panoptic controls now in place it is difficult to do otherwise, as the risk there includes total excommunication—that is, being banned from *eBay* altogether. There are many lessons to be learned from the way *eBay* manages its population.

The notion that communication media are social environments with complex relations should be an old trope to the reader at this point. If you're coming to this chapter first, however, consider the following: by virtue of their environmental properties, all communication media are ecological or systemic in nature. As with any natural ecological system, human-made media ecologies enable and constrain the entities that operate in or through them in various ways. They afford certain actions and behaviors, and limit or prohibit others according to the specifics of their design features, structure, and layout. A lake, a river basin, a floor plan, a skyscraper, a subway system, an interstate highway, a person riding down a bicycle path, or using a laptop computer to visit a favorite website—in every case, these systems carry particular biases, logics, and built-in predispositions that suggest and even prompt certain ways of experiencing, acting, and thinking about self and world.

Contrary to what many of the initial empirical studies of human/computer interaction seemed to suggest (cf. especially Bench-Capon and McEnery, 1989), people interact *with* computers—we do not merely interact *through* them unscathed. Or, as James Carey (1989) succinctly put it, technologies have "teleological insight." While not always singular, nor rigidly deterministic, every tool has certain *ends in mind*. Indeed, as one keen observer writing in the popular press noted around the same time Carey's *Communication as Culture* went to press: there seems to be a "supervisor inside" our computers (Kilborn, 1990).

A quick look at the massive technological apparatus that is *eBay* initially reveals a secure, stable, and highly efficient marketplace that has helped hundreds of millions of people around the world turn just about anything you can think of into extra cash—and nearly 40 billion dollars for *eBay.com*.[1] *eBay's* remarkable success as a commodity exchange hub is closely linked to its self-described civility. But upon closer inspection we find that this stable edifice conceals a rigid, if also subtle logic of action simmering just beneath the surface. The sublimation of control on *eBay* holds great consequence for social spheres traditionally conceived to be outside the realm of commerce, as the interfaces, systems, logics and functions that enable *eBayers* to do what they do continue to diffuse elsewhere.

In sum, this chapter is an investigation into the various ways *eBay's* environment encodes, describes, detects, and directs its users. After reviewing some of the commentary offered by research participants I suggest how the logics of action subtly proscribed on *eBay* might not simply be adaptive or advantageous to activity within this kind of online domain, but can be behaviorally prescriptive *off-line* as well. In between, I argue with respect to the conduct of social and political actors, how the personal thoughts, values and beliefs that generally function as reliable guides to human behavior begin to lose meaning as part of an operative definition of individual agency. We'll see how the formal features of *eBay's* virtual space—especially the user interface that centers around the feedback profile—engenders a reconfiguration of human perceptions regarding self and other that moves us from a human-human to human-object to even a kind of object-object configuration. While there is a deep intellectual history regarding such trends, a few key observers will suffice to frame the core issues here.

In 1651 Thomas Hobbes conceived of an orderly human society—the *Leviathan*—as a self-organizing system possessing a life and intelligence of its own. George Dyson, writing in *Darwin Among the Machines* (1997) draws out the contemporary significance of Hobbes' early notion of cybernetics or self-governing systems. Hobbes stated that "[h]uman society, taken as a whole, constituted a new form of life" and went on to suggest that "every joynt and member is moved to perform his duty" (in Dyson, 1997; p. 3). From this, Hobbes maintained, the *salus populi* or "people's safety" would be assured. Inspiring Jeremy Bentham's insights on the *panopticon* about a century later, Hobbes' thinking was that a healthy balance of internal and external surveillance—the policing of both self and other—would most reliably ensure peace and prosperity. Like the high-pitched détente sustained throughout the Cold War, the threat of effective penalty (if not mutually assured destruction) was seen to be a reliable instrument for maintaining modes of life.

Hobbes' leviathan, Bentham's panopticon, and later Foucault's (1973) analysis of sexuality and the categorization of medical conditions describe a history of emergent social control mechanisms. The fuller implications of these kinds of "socio-technical systems," described by Mumford, 1934; Heidegger, 1962; Ellul, 1964; Latour, 1988; and others, illustrate how the tandem technologies of *definition* and *categorization* work toward the establishment of subtle, but nonetheless highly effective behavioral norms that can be quite tyrannical in their capacity to direct and constrain the thoughts and actions of individuals and collectives alike. As Dyson puts it, the "diffuse intelligence of the Web and all the nodes of processors have more in common with Hobbes' leviathan than with the localized artificial intelligence, or AI, that has now been promised for fifty years" (Dyson, 1997; p.7). While Dyson refers primarily to small servers and personal computers functioning as part of a diffuse, global data bank-qua-intelligence, his analysis of modern digital computer networks ultimately extends to include human users working in various degrees of symbiotic relation with their machines.

SOME METHODS AND QUESTIONS

As with the previous chapter, the present analysis was prompted by some of my own experiences in Digination. I sailed into *eBay* for the first time in 1999, with my first full-time teaching salary and a serious buying compulsion that lasted for about one year. My first big purchase was actually not conse-crated through *eBay,* however. I had never paid more than $600 for a car and figured now was the time to spend some money. Folks didn't typically sell cars on *eBay* back then anyway. For one, *eBay Motors* did not yet exist, and PayPal had just incorporated a year earlier so I had yet to hear about all that. The car I located was on a specialized VW enthusiast site. Sort of like the way *Craigslist* handles cars these days. I drove a few hours and paid for my Corrado in cash. The guy sent me a few low-res pics but it really was, so to speak, "sight unseen." And it was an exhilarating experience.

Once back at my computer desk and surfing *eBay* I fed a penchant for original Star Trek episodes and artifacts, as well as vintage snowboards that kept me up late more than a few nights in an effort to avoid the "snipe" (the surprise, last-second bid from another overzealous bidder, or their automated "bidding bot"/*sniping* software). These personal experiences spurred my interest in the social and communicative significance of *eBay.*

What follows is an ethnographic inquiry that echoes some aspects of the Mohawk e-mail study. I also performed content analyses of user-generated commentary obtained via participant observation and depth interviews. I personally burned out on *eBay* after spending $1100 on three different transactions one sleepless weekend a few years back. After eight years or so operating in the *eBay* environment, and having completed perhaps two dozen transactions during that time, today I am just a sporadic lurker. But I still get the sense that something new and interesting is going on on *eBay.* And it might be propagating throughout our Digination.

The core questions that guided the investigation include: How is the self/ person negotiated and maintained on *eBay*? What is the nature of the relation-ship to specific others and the collective? What does successful interaction look like and how is it structured? How are "violations" dealt with? And finally, how might behaviors adapted to this online environment contribute to an alteration in the way people think about themselves, each other, and the wider society both on- and off-line?

With the aid of ten research participants, we see some of the ways personal thoughts, beliefs, and values seem to melt away as impractical, nonfunctional, even disadvantageous concerns. And yet these are issues that often form the central features of analysis in contemporary psychological studies of social interaction. It was not uncommon, in fact, for my research participants to report that they were somehow compelled to act as they did. "I had to," "I didn't really have any choice," "this is just what you do," "sellers need to be polite" and "you better not mess around" were common retorts to questions of intention and action.

Again, McLuhan's "laws of media" go far in helping us understand this new environment. Recall that the approach characterizes and even makes tentative predictions with regard to how any technology potentially enhances, reverses, retrieves from the past, and obsolesces particular aspects of human experience (cf. McLuhan and McLuhan, 1988). The tetrad reveals how, through a collection of procedures and rituals that become necessary in that setting, the structurally determined techno-social construction of self results in a fairly specific kind of inhabitant. In other words, *eBay* frames, formats, and otherwise presents not only its products, but also its users in systematic ways. Indeed, we'll discover that it is the users who become the product-qua-content on *eBay*.

The techno-social formatting that takes place on *eBay* has an analog in something Joshua Meyrowitz (1985) dubbed the "staging contingencies" of television. Like the commercial television context, *eBay* pre-figures and often reliably predicts the kind of individuals that will surface in its environment. In so doing, the site prompts something I've dubbed "ostensibly civil behaviors" in a patterned and systematic way. The behavior prompted and channeled by the *eBay* environment constitutes a still nascent but already robust normative code that may be representative of an emerging norm on the Internet itself. Nicholas Carr (2008) is certain these issues should concern anyone who uses the Net: "[n]ever has a communications system played so many roles in our lives—or exerted such broad influence over our thoughts—as the Internet does today. Yet, for all that's been written about the Net, there's been little consideration of how, exactly, it's reprogramming us. The Net's intellectual ethic remains obscure."

I'm with Carr on this. From the tech expert to the kid bidding on *Bakugan*, it remains difficult to see what's really going on here. And especially where the ethics and morality of this emerging code is concerned, there are a number of pressing questions that remain. But while this chapter is an inquiry into social behavior online, it is not a moral analysis, or even an evaluation of individual human intentions. People trade on *eBay* because they find it beneficial or rewarding for a variety of reasons. Something on the order of 1.5 million Americans currently make all or a substantial portion of their living from auction revenues—with an increasing number claiming all (*employment.families.com; ACNielsen International Research*). These kinds of statistics sometimes seem questionable, so I looked into the details a bit further. What I discovered is that most of these folks are still just scraping by. Nonetheless, most hold great personal stock in the viability of the whole enterprise, and are proud to be star-sporting eBayers.

There is also a segment of the *eBay* population that goes about things a bit more casually to sustain a small side income, and another cohort that just makes a few extra dollars in the course of enjoying a hobby or favorite pastime (I suppose that's where I fell in with my Star Trek and snowboard excursions). Still another group simply gets a thrill out of the whole process—whatever the item, buyer or seller. Finally, there is a very small population of dispersed grifters, charlatans, and the flat-out frauds. I got stung by one of these a few years back and know that if someone puts their mind to it there are ways to avoid paying for items received from an unscrupulous seller. There are also successful ways to avoid sending items that have even been paid for through the *Pay Pal* filter. But *eBay's* security mechanism has become so robust since then that it is increasingly difficult to evade sanction. In any case, these are some of the primary motivators and logics of action on the site. And we'll find that certain descriptors like "grifter" and "charlatan" become uneasy, imprecise, hypocritical sorts of labels.

THE MECHANICS AND MEDIA LOGIC OF *EBAY*

eBay hosts peer-to-peer interaction sustained through a highly formalized feedback mechanism centered around the exchange of goods that is bound, almost gravitational-like, by the network effect of hundreds of millions of daily users. To be sure, "[n]ever before has there been a market with such abundant dimensions" (*Economist,* 2005). At first pass, these processes appear to be working symbiotically with and through *eBay* users who dutifully and often enthusiastically sustain a self-regulating social control system that explicitly models and patterns successful behavior, thereby making transgressions a very rare occurrence indeed. However, we find that this patterning fosters an eccentric and highly self-conscious kind of social and political consciousness that itself engenders a "culture of consent" which may be particularly susceptible to hegemonic control. *eBay's* feedback system incidentally keeps tabs on virtually all activity that occurs in reference to auctions, payments, the shipping and handling of goods, the fulfillment of transactions, as well as the periodic breaches of each.

Through the feedback system and the item posting process users end up serving another function. They are *organic capacitors and transistors* that do the lion's share of work for *eBay* in terms of advertising, promotion, quality control, security, and the distribution and circulation of commodities and capital. While *eBay* pedals virtually any kind of material item, it is the social and cultural capital in the form of users' identities that is of particular interest here. *eBay* users (or *eBayers*) are the primary dispensers of the cultural capital that created and now maintains the *eBay* persona—the corporation's most precious commodity. On *eBay*, a global intelligence with a collective aim is now distributed among a number fast approaching three hundred million regular users worldwide. *Everyday eBay* (2006) contains a broad collection of analyses focusing on the online auction house as an important socio-cultural phenomenon. In the introduction to the volume, it is pointed out that

a crucial component of eBay's success, both economic and cultural, it its organization of the site as a series of stages allowing sellers to design, perform, and sell memorable experiences, thematically linked goods, for which purchasers are willing to pay a premium. These performances, and the willingness to pay, exceed commonsense understandings of eBay as a giant garage sale or the old-fashioned auction house writ virtual (p.1).

Extending ideas developed in two of that book's most media-centric chapters, the current chapter reveals how the formal characteristics of *eBay* do much more than merely *allow* these things to happen.

A user's feedback rating is conspicuously displayed at the top of their profile page, visible to any visitor with a single click. As we'll learn through the commentary provided by several participants, all felt in their own ways that their *eBay* space, and their reputations indexed there, have come to mean something more than they might have imagined—and probably more than any other *place* online.[2] One side effect of this indexing is that an individual user's digital identity becomes a highly constructed and tightly controlled commodity inextricably bound up in the *eBay* brand. The development and eventual achievement of a good reputation on *eBay* becomes a transparent, legible, highly intentional, and strategic object to be designed and manipulated by users.

Social interaction on *eBay*, already displaced in time and space by virtue of its highly prescribed, asynchronous interface mechanism, is further loosed of its moorings to the here-and-now of social life as *eBayers* seem most interested in closing the auction or sale, getting on to the next, and enhancing their feedback statistics along the way.

Of course this is just good business. And it is nothing new. The merchants of ancient Cairo, Athens, Istanbul, and countless other trade hubs around the world also endeavored to be efficient and therefore sought positive recognition as reliable sellers in the open marketplace. To do otherwise was to risk being eclipsed by the competition across the way—or worse, banished from the marketplace altogether. In the process, these cities gained distinction and prosperity in step with the workings of their markets. Though the incredible pace and scale enabled by digital commodity exchange is something new, it is not that we are seeing a wholly new kind of activity taking shape on *eBay*. Instead, we're seeing a new kind of personality, a new sub-species of social being emerging—one adapted to the new forms, pace, and scale of social interaction that typifies much of our relationships with and through computer technology. *eBay's* rigidly structured, primarily text-based interface that provides most of the social context for interaction is highly consequential in its capacity to realign values and perceptions of what constitutes successful actions and, by extension, elicits positive and locally meaningful (read *functional*) social behavior from its users.

Research participants in this study offer insights that suggest how personal thoughts and private outlooks are rendered peripheral, even superfluous concerns in the world of *eBay* because they tend to be non-functional, even inconsequential to the way people behave there. Given this, we should wonder if interaction patterns that are functionally beneficial in the *eBay* domain translate well to a sustainable way of being off-line, in the "real world" so to speak. Of course, the consequentiality of acts and deeds in *eBay* has been impacting the off-line world for 15 years now, as more people set up accounts every day and feel the very real consequences of money orders, checks, electronic deposits, and stock values. There is more to the story, however, and it concerns a psychic shift now in process. So opposed to an inquiry in individual intentions, we need to draw out the nature of these side effects and the unintended social and psychological consequences of dwelling and acting in a "place" *eBay*.

RETURN OF THE *MODERN*: CODIFYING STABILITY, SECURITY, AND SOCIAL CONTROL

The most significant observation made in *EveryDay eBay* is one that turns much of the extant cyber-identity and cyber-politics literature on its head: the "reassertion of a fixed, stable, and very modern identity formation" (p.5).

eBay represents a virtual commons where the postmodern analysis of human identity as an adaptive, fluid process finely tuned to the contingencies of the social environment becomes counterproductive. To the contrary, in order to function successfully as a citizen of *eBay*, one needs to act out an almost mythical conception of identity—one that is hyper-stable, consistent, and totally explicit. "Apparent transparency" is the name of the game.

As opposed to being representative of an actual person in the flesh-and-blood social world, these fixed, hyper-stable selves give way to highly predictable behaviors and most resemble the stereotypic door-to-door salesman or overzealously tip-minded wait person at an upscale eatery. More darkly perhaps, the model *eBay* citizen conjures the Machiavellian caricature of a calculating political candidate, social charlatan, or consummate grifter, so a Nietzschian genealogical analysis of morals could be fruitful here as well. Ideas, objects, or people—there is no functional distinction when it comes to closing the deal—all is quantified and rendered a utility on *eBay*.[3]

Individuals cultivate the image of being consistent in both thought and deed, a stable and predictable actor in the world of online exchange. As such, all stands on the verge of lapsing into artifice as the individual finds themselves acting, even *acting out* in a kind of perpetual campaign mode in their interactions with others. The human relationship—ideally based on something akin to Buber's (1977) *I-thou*/subject-subject relationship, or Schutz's (1972) *We* relationship—reduces to a mechanistic, functional, subject-object relation. In this unfolding, task-oriented, *I-It* dynamic, an *eBayer's* social consciousness moves systematically from *Thou* and *We* to *They* and *other*, as virtually all interaction on the site becomes a means to an end, a subject-object confrontation. It is even arguable that a peculiar kind of *it/it* relation develops on *eBay*. That is, individuals can also come to see themselves primarily through an objective lens (Mead, 1934). Considering some initial participant commentary, we see how certain features of the *eBay* interface generally supports and sustains this mode of being and seeing. An account provided by one *eBayer*, an Indian-American graduate student in his mid-twenties (here dubbed *Adin*) provides the first example of this perceptual shifting vis-à-vis others in the *eBay* domain.

> I wanted to make some money . . . you know, buy a bunch of clearance items at *Target* and *Walmart* and such, then sell them for five or so dollars below standard market price on *eBay*. I knew I wouldn't get rich . . . but, you know, to get my [feedback] rating up. It was all good until around Christmas, 2004. I sold a *Game Boy* cartridge to a guy in Colorado, shipped it via media mail. After four weeks, he e-mails me and says it didn't arrive. I tell him to give it another week, to let me know if it still hasn't arrived, and then I'll refund his money . . . because I can say that my rating was more important than some $15 game!

So he gets back to me and says it still hadn't arrived. Another e-mail round and I find out he eventually got it—and this is exactly why I was waiting! But it's crushed. USPS is at fault, not me. The problem is that he then lets me know that he had already left me negative feedback. So I think . . . ok, he's being honest. I didn't give him negative feedback in return . . . he was being honest. But I was irritated because it took me down to ninety-eight percent. I was thinking—do I want to use up this opportunity to make a claim? I was looking into the future. I was thinking *eBay* will be around for a while. I see it [my feedback score] like my credit history. Even though I accepted it as one of the risks you take using *eBay*, it still bugged me. I thought about leaving that guy negative feedback for almost a year.

Adin's example vivifies the manner in which *eBay's* interface prompts users to adopt a new awareness with respect to the relationship between their current activity and their past and future behaviors, and simultaneously sublimates the feeling to act in what might otherwise be perfectly appropriate ways. Adin felt at odds with himself in a manner he later described as "uncomfortable," "unfocused," even "unnatural." Adin knows well the contours and general ins and outs of *eBay,* and he wanted to preserve certain, finite rights afforded him. It was standard at the time of this writing for *eBay* users to be allowed three claims that include the reimbursement of lost or missing funds. *eBay*, in other words, will *take your word for it* on three occasions. On the surface, this all corresponds perfectly with the standard "three strikes" rule applied in many "real life" situations where there is an effort by some kind of authority to enforce appropriate behavior. The equation also seems to square nicely with one of *eBay's* five "fundamental values." As part of their effort to promote a communitarian ethic that, hopefully, creates an idealized sort of "small town" civility, *eBay* conspicuously proclaims in its *five beliefs* that:

1. We believe people are basically good.
2. We believe everyone has something to contribute.
3. We believe that an honest, open environment can bring out the best in people.
4. We recognize and respect everyone as a unique individual.
5. We encourage you to treat others the way you want to be treated.

We believe . . . The rhetoric of hope is alive and well on *eBay*. Like Massachusetts governor Deval Patrick's winning slogan: *"Together We Can,"* or Barack Obama's employment of a similar ethos with his message of hope through change, something is being tapped here that may be an essential part of the human psyche—a whispered promise pointing to that unwritten contract of civilization. The notion of a better future has always been a compelling story for *homo narrans*, and *eBay*, via its feedback mechanism and online forum, crystallizes the dream. But is it a dream perpetually deferred? Dwelling in this corner of Digination, can one ever be made to feel satisfied with their social position? In a place where virtually anything is attainable, where one person's detritus morphs into countless others' desiderata, how do we know when enough is enough? As Jarrett (in Hillis et al. 2006) puts it, "[t]hroughout the entire site, *eBay's* representation of itself minimizes its own authority" (p.112). Such veiled injunctions create what amounts to a procedural dictate, a social edict to be good and proper (and ideally successful and wealthy too). And yet, "*eBay's* ability to make unilateral decisions, such as releasing consumer information to law enforcement officials, remains intact" (ibid).[4]

> A liberal government works "at a distance," utilizing an array of technologies of the self—technologies that produce subjects already geared toward the dictates of liberal rule—rather than forcible impositions of its will. This hegemonizing impetus places a premium on seemingly noncoercive mechanisms of internal discipline, sites of normative judgment, and expertise through which 'correct' practice is inculcated and normalized within the psychology of the individual citizen (Jarrett, 2006; 109).

Rosanne Stone (1995) offered this basic critique early on in the Internet age by describing a legible body or subject that comes to represent the standard denizen of the Net. Stone predicted what has come to pass: the emergence of "fiduciary subjects" that have value primarily due to their ability to engender public confidence. Once again, digital communication technologies did not bring this situation into being. Some of the first record keeping devices included marked bones, rocks, and animal skins to tabulate the number of kills a hunter amassed in a season. Notched and cleaved wooden sticks were eventually mated back together to ensure proper repayments on loans to and from the king. Human beings have, in other words, been developing means of extending, tracking, and tagging the weight and measure of personhood beyond the body for thousands of years.

Today, however, we establish, gain and lose credibility through a host of methods and means. These virtual and physical mechanisms include but are not limited to: toll fare transponders, RFID tags, merchant discount cards and point-of purchase technologies, credit cards and credit reports, census data, tax statements and filings, legal documents, street and e-mail addresses, personal webpages, online profiles, atom- and electron-based catalogues, lists, photo albums and journal entries, telephone numbers, high school and college diplomas, insurance policies, academic progress reports, and medical histories. In the midst of this our hyper-legible digital selves become devices that can certainly liberate us in unprecedented ways by reducing the time it takes to get on and off the interstate, or shop or correspond with friends and family, or determine just what it is we did during that trip to Vegas three years ago. But they can also prove to be a useful means of "articulating the political ties between the discursive terrain one inhabits, and one's sense of self (Nideffer, 1996; 484).

THE TYRANNY OF TEXT: SUBVERTING THE SUBJECTIVE

Adin's experience illustrates the substantial causal power of the textual format symbolizing most of the action on *eBay*. The persistent, public, visual quantifications and one-line qualitative comments forming the backbone of the *eBay* user interface configures and constitutes the operative components of user profiles.

In *Empire and Communications* (1950) Harold Innis undertakes a fine-grained historical analysis charting the social and political effects forming the wake of a progressive build-up of fixed cultural records in Mesopotamian and ancient Greek and Roman society. These durable physical records represented trade shipments, personal and collective holdings, the scale and equipment stores of armies, livestock ledgers, and the like. Since Innis, we have seen the advent and broad dissemination of persistent records composed not of atoms, but electrons, and they now represent everything from the genetic sequences of parasites to the economic status of nations. However it is in that middle range of analysis, between the microscopic and the transnational— that is, along the human scale—that these processes and practices become highly consequential to the way people think about themselves, each other, and the wider societies they inhabit.[5]

While Heraclitus and Parmenides toyed, respectively, with fluid and stable theories of self in ancient Greece, it was not until the mid-sixteenth century, when Michel de Montaigne suggested that the "inconstancy of our actions" was a central part—if also at times an inconvenient piece—of what it means to be human. But if Montaigne's important essay never gathered much of a following in his own time, by most accounts today, he (like Heraclitus) seems to have been most on track. Indeed, Montaigne's postmodern theory of identity had to wait four hundred years before a broader awareness of the socially orienting, context-dependent and often very fluid aspect of the human self gained prominence. With Burke's and Goffman's *actors* then taking their cues from the social situation, the apparent needs of the audience, and the practical purposes at hand. Social constructionism is born. Yet this recognition of a more flexible, situational self described so well in Goffman's accounts (1959, 1967) is losing traction in the age of always on, always visible, always working digital identities where fluidity and change is seen as suspect—a discernible character flaw writ large for the world to see.[6] And there is good reason for this. "When identity was defined as unitary and solid it was relatively easy to recognize and censure deviation from a norm" (Turkle, 1995; p.261). Contrary to Sherry Turkle's suggestion that "[a] more fluid sense of self" seemed an inevitable consequence of life online, that would allow "a greater capacity for acknowledging diversity" (ibid), more and more of these digital spaces we are occupying do not allow the naturally fluid aspects of our selves to emerge—aspects that have, arguably, always been closely linked to the immediate social contexts we inhabit.

One notable feature of the self-inscriptions on *eBay* is their openness to mass exposure. But in this *eBay* is also not unique. *MySpace, Facebook, YouTube* all represent test beds that are revealing some of the psychological, social and political reverberations this strange new kind of public legibility plays in our lives. While few *eBayers,* at the time of this writing, utilize the personal photo option, there is a slow trend to take this up as a reliable strategy to ensure stable business on the site. Other sites, of course, are explicitly geared toward "full exposure." Personalization, and "putting a face on things," is generally an effective way of winning interest and favor in most electronically mediated contexts. *Match.com, eHarmony,* and virtually all of the successful online dating sites, for instance, learned quickly the economic (and for their members the practical) benefits of the photo image as a way to minimize deceit and increase control of the socialization process.

Adin described a kind of hyper self-consciousness forged during his interaction on *eBay*. Even though one's personal profile has to be intentionally logged into upon entering the domain, other users' short, verbal descriptions of him, and the numbers *eBay's* computers align with those descriptions, always seemed to be, as he said, "in your face." This persistent data edifice looms as a constant reminder—something like a *personal social capital credit report* for all to see. In fact, we'll see that this is nearly the way Adin himself described his *eBay* profile (some of which is shown here):

model great, shipment fast and safe, thanks for everything	Buyer cur (66 ★)	Jul-27-06 19:37	230008810468
Good buyer, prompt payment, highly recommended.	Seller ba (3112 ★)	Jul-16-06 17:04	140007758453
Top bloke to deal with, very friendly and fast, smooth transaction AAA+++	Buyer sq (154 ☆)	Jul-05-06 15:52	6066441206
Excellent transaction, fast postage and well packed, very happy.A+++	Buyer sq (154 ☆)	Jul-05-06 15:47	6066442791
paid same day as invoiced and EXCELLENTcommunications - HIGHLY recommended !!!!	Seller sel (1086 ★)	Jun-25-06 20:01	8063404744
Pleasure to deal with. Prompt and Professional. Recommended!	Seller au (44599 ☆)	Jun-15-06 21:27	6060162010

This profile information seems a bit like the imagined small talk in a mid-twentieth-century Rockwell-esque diner or grocery store. But these words are not ephemeral in the way spoken words are. They do not fade over time and space. To the contrary, when several of Adin's deals went sour, the conspicuous legibility of his online persona (most notably, the visibility of his feedback rating) frustrated him for more than a year. Adin was, as he put it, "constantly reminded of his [the other *eBayer's*] incompetence" in navigating the exchange process, as well as his own impatience and impertinence via e-mail while trying to rectify things and salvage the transaction. But above all, for Adin, this was a matter of maintaining his good name: "I can't get that 2 percent back! It would probably take me 10,000 transactions to get those numbers back up near 100." Up near 100?! I wondered, is 98 not near 100?

Of course, it could simply be that these concerns are the personality quirks of one research participant—the obsessive, compulsive twinges of a perfectionist. The problem with such an assessment, however, becomes clear when we step back and wonder: who wouldn't want a flawless reputation in the "real" world? That possibility is the stuff of dreams: the good and helpful neighbor, the model citizen, the popular hero. And yet, without making any essentialist claims about human identity, a key difference in the way great reputations are won and lost on *eBay* is the near total lack of usable subjective perception (and with it, spontaneous action) the site allows. There are multiple ways *eBay* tries to turn subjective opinion into objective assessment before the sale or close of auction: the photo gallery, the disproportionately large space allocated for item descriptions, a quite thorough (though easily missed) set of guidelines for posting feedback, etc. It is during the reception and provision of feedback commentary when one often must bite their tongue so as to appear affable and safe to do business with. Building on Adin's perspective, some of my own experiences offer further insight into this phenomenon.

STUMBLING AROUND IN THE *GENERALIZED NOWHERE*

Much like Adin, I tended to be on the lookout in the off-line world for good deals . . . and always with *eBay* at the back of my mind. In the late spring of 2005, I ran across a coveted *Ride Prophet* snowboard in a sporting goods store returning from a great mid-week day on the slopes. The item, on consignment, was in near-mint condition. There was barely a scratch on it. I later discovered that the board had never been owned but was a twice-used demo at a local high-end ski shop. The price tag: *$150*. Standard in my opening volley to any dicker, I asked the guy behind the register if he could knock that down to $125 or so. Without hesitation he said "sure," and added, "that's a great board, last year's model . . . I can't believe it's still here." I agreed, and realized I probably could have asked him to take another $25 off without a fight.

Truth is, not many people ride boards like that—a bit long at 168 centimeters—but I knew I could fish out someone online who did. I plunked down my debit card thinking the graphics weren't really my style anyway. But it was certainly the nicest board I had ever owned. Given this, and as the closest semantic equivalent available when posting an item on *eBay* was "excellent," I selected that option without a second thought, and followed this with considerable detail about the rare condition of the board in the written description section. I was not then aware that *eBay* had its own rating advice. And indeed, this is where my trouble began.

The auction itself was an exercise in perfection. I felt sure enough about my find to start the bidding at $100, with no reserve. If my prediction was that I might even double my money, I knew I would at least make it back if things didn't go so well. I checked in periodically over the next week to see how things were going. 24 hours before the end of auction, the highest bid was $189. As luck—or the market—would have it, the winning bid ended up being $325. I nearly tripled my investment. Like Adin's fellow gaming enthusiast, my buyer lived in Colorado, and the estimate of $20 I had posted came close to the actual cost of shipping via UPS ($26 for second day air). I sent off the package and eagerly awaited my positive feedback. It came soon enough (first via private email) but with a twist. The initial message is followed here with my immediate response.

it arrived today, thanks, I appreciate it. great board, but if you used that board only twice, im the friggen pope . . . its rusted and has been nicked 10 or 20 times . . . ill use it a 100 times this year though . . . we were open to 4th of july

———————

Hi,

Sorry you don't think the description was up to snuff. I probably should have gotten the bottom ground. But I tell you that those are considered superficial blemishes on the bottom + certainly on topsheet where I'm from. I'm realizing from your e-mail that there may be a different notion of "condition" here on the east coast than out west. I did in fact only ride the board on two weekend trips in NH and VT. We don't have the deep cover that you guys are accustomed to. Again, I am sorry about this misunderstanding. Could I refund some of your money—and perhaps pay for a stone grind to lessen your dissatisfaction?

let me know -r

I offered my angle on what may have occurred—namely, a difference in experience which led to a difference in description. But this did not deter my buyer from stinging me later that week with a subtle but nonetheless consequential dig that would become a permanent piece of my official *eBay* profile. In just one line it simultaneously documents my efficiency and what he perceived to be my deception leading up to auction. Rhetorically, this all but neutralizes the two positive remarks offered at the beginning of the comment:

> Good seller, fast shipper, can be a bit liberal on condition though

And there it was fort all to see: "Can be a bit liberal on condition." Now, if I was a serious *eBayer* this would be the part that really sticks in my craw. The guy didn't know me from Adam, and we certainly had never done business before, yet the statement suggests something enduring about my personality—that I have a tendency to embellish. It's funny when I think about it because my 6-year-old son would agree. Dad tells stories. But on *eBay* I was so careful to tell it like it was. Unfortunately, telling it like it is (or as one thinks it is) on *eBay* can be problematic. The rhetorical twists and subtle allusions to hidden dissatisfactions that make up so much of the feedback there often reads like a catalogue of perverse epistolary exchanges full of veiled threats and subtle stabs between long-time foes.

Unfortunately, this potentially educational rhetorical exercise (stylistically at least) is typically limited to one-liners. It is an allocation determined by the rigid limitations of the feedback interface. Just below the syntax there often churns a maelstrom of dissatisfaction and deceit. Contrary to the argument exquisitely laid out by Turkle in 1995, there is nothing therapeutic about this increasingly common aspect of *life on the screen*. And it is notable that Turkle retracts much of her earlier thesis regarding the positive potential of Digination in *Simulation and its Discontents* (2009). The fact is that these *eBay* experiences testify to a sentiment long ago offered in the epilogue to Neil Postman's prescient *Amusing Ourselves to Death*:

> What Huxley teaches us is that in the age of advanced technology, spiritual devastation is more likely to come from an enemy with a smiling face than from one whose countenance exudes suspicion and hate. In the Huxleyan prophecy, Big Brother does not watch us, by his choice. We watch him by ours. There is no need for wardens or gates, or Ministries of Truth (1985; 155).

There is a closely related passage written by William Gibson in his characterization of the *Press* in Singapore in the 1990s: "[t]his ceaseless boosterism, in the service of order, health, prosperity . . . quickly induces a species of low-key Orwellian dread (the feeling that Big Brother is coming at you from behind a happy face does nothing to alleviate this)" (Gibson, 1993). A similar boosterism reduces virtually all of the commentary on *eBay* to a kind of virtual lip service that nonetheless has very real consequences. Huxleyan or Orwellian—neither seems particularly appealing—as *eBayers* become their own (and each other's) wardens, gates and ministers.

Some grassroots efforts critiquing the overly rigid, limiting aspects of the interface (or more often the murky policies of *eBay's* sister site *PayPal*) have smoldered and continue to catch fire in various corners of the Web. But these detractions tend to be little more than distractions, momentary disturbances, digital blips on the serene surface that are all but ignored by the larger *eBay* population. Like Google and Microsoft before, the inexorable network effect bolstering *eBay* activity seems to have little patience and leaves even less communicative space for criticisms and concerns. The embedded logic of action is incredibly powerful in this new environment. But while not particularly subtle once you get the gist of it, it is not clearly deterministic either. The irony is that it is hard not to take part in the gambit. Of the ten research participants involved in this study, the ones I've dubbed Adin, Tom, Andy, Dan, Jen, and Tracey reported direct if ultimately unwilling participation in these processes. If not admitting to overtly conscious activity, these users did describe a kind of enabling behavior as they tended not to report negative activity where it would have normally seemed appropriate. Such outcomes seem based most generally on fear of reprisal or the impression that complaining simply isn't worth the effort.

Take Dan, a thirty-something car enthusiast living in suburban Boston, who bought a used performance part for his vehicle from a seller in Montreal. "I was so excited about the purchase that I forgot to send off the standard positive feedback." Dan eagerly described to me a flawless transaction where the price was right. He had intended to add a row of stars and other accolades to his seller's profile, but quite simply forgot to act out this needed moment of civility. That is to say, no ill intent was involved: "I'm just sometimes a bit scatter-brained with a sketchy memory." Two weeks later, when he happened to be perusing *eBay* for nothing in particular, Dan found this message listed as a "follow-up" below the positive response the seller had initially left for him the day the funds were transferred:

Smooth and fast transaction, thanks.
Follow-up by Cjxxxxxxx; tried to contact him after to see if everything alright, got no news from him.

As with my own snowboard example, the implied jab here too is anything but subtle. In the context of *eBay* this is akin to being insulted when someone simply forgets to say "thank you," but with the additional suggestion that he (the seller) did all he could to determine if everything was ok.

The seller ("Cj") certainly appears to have been distressed by the fact that Dan did not leave him the official (and positive) feedback he expected. As soon as he saw this Dan replied via email to indicate that everything was fine with the delivery and that the component fit and worked perfectly. What happened next caught Dan by surprise. The seller e-mailed Dan back and asked him to write the gist of the email in his feedback profile. In other words, the seller's inquiry did not appear to be motivated by any real concern for Dan's own satisfaction. Rather, Cj's interest was that this transaction be properly detailed and his feedback score enhanced for all to see. He needed this deed made public, at least for all *eBayers* who happened to come upon his items and profile. I refer to this as the *Facebook effect* in my classes, for it is a phenomenon that is ubiquitous with young people today: the need for "positive" exposure . . . of some sort.

This kind of motivation seems perverse at first glance. However, upon further reflection most parents will admit to training their children to compliment and apologize, to say thank you and please even when they sometimes don't really mean it. If this somewhat propagandistic aspect of our behavior has always been a necessary layer of surface sociality (and it certainly typifies business transactions in most parts), it becomes intentional, procedural, even institutionalized in the *eBay* polis. The car part seller's desire to be seen in positive light by all is hardly an isolated instance of the kind of overt expectation that has emerged on *eBay* in this regard. Nothing goes uncounted and, ideally, all is appended to the record. To be sure, the data edifice is insatiable. For instance, at one point, after nearly four months of being away from *eBay*, I found the following very pleasant notification in my email:

> Congratulations! You've achieved a feedback rating of 10 and that means you've earned a Yellow Star next to your eBay user name.
> We want you to know how much we value you as a member of the eBay community. You are our foundation—we wouldn't be here without people like you.
> Again, congratulations on your success, and keep shooting for the stars!

A pleasant enough note, but this prompted me to second-guess myself to determine that I had, in fact, not been active on *eBay* more recently than originally thought. Of course, the communiqué is just an automated response that is triggered after a lengthy absence. The message conjured some fond memories of taking piano lessons as a kid and receiving stars for successfully negotiating a couple of Mozart's more squirrely minuets. But this felt more like a half-hearted effort to get me back in the auction game as soon as possible. Like a lot of advertisements, the rhetoric is at once pragmatic and patronizing. Then, another month goes by and this arrives via email:

> As an eBay member, you get publishing space to tell the community what makes you . . . *you*. eBay *My World* lets you share all your favorite things, what you sell, what you buy, and so much more. Click the Add Content button to keep personalizing your page and add new modules such as guest book.

With *The Mechanical Bride* (1951) Marshall McLuhan revealed his fascination in subliminal messaging—messages that operate on a preconscious or pre-logical level. As McLuhan put it, "any ad consciously attended to is comical. Ads are not meant for conscious consumption. They are intended as subliminal pills for the subconscious in order to exercise an hypnotic spell" (1964, 218). McLuhan's insights illuminate the ceremonial nature of communication on *eBay*. Indeed, the commentary from Adin, Dan, and other *eBayers* suggests that the feedback profile and fora carry little real informational value. If anything, with what really amounts to a binary assessment (given the negative valence *eBay* itself attaches to neutral feedback), most of the communication on the site serves more of a self-referential ritual function that again points to James Carey's (1989) assessment of so much modern communicative practice. Though part of this also stems from the way sellers in particular must attend to their behavior.

During our interviews Adin returned often to the manner in which the tables are turned between buyer and seller: "[f]or me, as seller, it's the opposite of buyer beware on *eBay*. I'd say it's *seller beware*. My responsibility is to protect the buyer from the 'buyer beware' idea as much as possible. Because they're buying it sight unseen." Adin's assessment seems on the mark. And the reason for this is clear. As Andy, a film collector and dealer with an admittedly "checkered" past on the site, remarks: "if you tarnish your reputation as a seller it can really come back to haunt you." Despite the ability to post multiple high-resolution images of items for auction or sale, Adin and Andy reveal a very serious concern for their buyer's stock of information before the transaction has taken place. However, upon further probing, we come to find that it is a peculiar kind of concern as Adin himself admits that a lot of the worry really does stem from his own compulsion to "stay as close to 100 percent as possible." Adin describes another scenario in an effort to explain this: "There are two versions of the fifth generation iPod. The thing is they look identical. So I could see where a seller might legitimately believe and even get others to think they think they are selling the '5.1' when they aren't really."

Notable here is the way a hyper-objective outlook—a kind of collective apprehension—is assumed. One better know exactly what they are doing in this domain. A broad and fairly deep awareness of the market (whichever one happens to be dabbling in) should be in place lest the *eBayer* risk looking like a scofflaw. Much like I ended up doing in the snowboard example, Adin ultimately provided a good rating for his questionable buyer. He bent to the benefit of the doubt. However, unlike Adin, I was never able to secure the full participation of my new friend in Colorado. Despite my honest appraisal of the item, and despite the vitriol seeping out of the e-mail he sent me during our post-transaction exchange, he fashioned an ambiguous enough piece of positive feedback to help ensure a good rating for himself as a future buyer and seller on *eBay* who is even-keeled, level headed, and easy to do business with.

These examples suggest that the users' individual will, and their inner thoughts and opinions are being systematically effaced by the formal logic of the system. Ironically, given *eBay's* own talk about itself, there is little civility to be found in these examples. A thinly veiled Hobbesian dark-world with most inhabitants acting in their own self-interest percolates just beneath the surface. And yet most of it, as another participant put it, reads "like the ravings of a perpetually happy person on Prozac." To be sure, the way *eBay* is structured, the site ends up functioning a bit like Adam Smith's free market on steroids, with a sprinkle of anti-depressant thrown in for good measure. The thing is, from a practical standpoint at least, it works very well for all parties who maintain active membership.

But strangers and itinerants beware. As Tracey, a college professor and music and movie fanatic suggests, "to be accepted on *eBay* you have to be more than a regular. You have to sort of live there." And the proof is in the pudding. Jarrett, writing in *Everyday eBay*, notes how "this is clearly a more liberalized Panopticon than Bentham's model, but not only in that it denies the centralization of power in an appointed elite. It is also more liberal in that it effects its disciplining, not in the form of coercion by an external force, but disguised as an exercise of free will and autonomous, responsibilized activity," (in Hillis et al. 2006; p. 116–117). The air of autonomy and responsible activity Jarrett notes is still apparent in most corners of *eBay*. In an e-mail extending our conversations, Adin explains some of this further with the different valuations he places on items won:

> I bought [sic] some counterfeit games from a guy in Asia w/o realizing it once. Branding was important to me. The imprint on the casing read "Nintondo." So I asked him to refund my money or I would report him to *eBay* about these fraudulent products. He said return it and he'd give me my money back. So another couple of weeks went by and I was refunded, and you know, he forgot about the games. He went out of business a week later—dead account.

It didn't matter that the "fake" versions of the games worked perfectly. After he discovered they were generic copies, Adin had to admit that he was making a purely subjective, even elitist aesthetic distinction. As in my experience with the snowboard sale, might Adin's seller have been thinking from a different *cultural* angle? For Adin, function clearly holds a subordinate relationship to form, and he was universalizing his own perception of that relation onto his seller who may in turn have conceived of an inverted (more pragmatic?) relation between form and function as a matter of personal, even culture valuation. This suggests that the unique perspectives of interpretive communities can be effectively neutralized on *eBay*, where a subject position grounded in the here-and-now becomes a liability. In its place a kind of homogenous global outlook is expected and assumed that has one eye on developing feedback commentary and the other on *eBay's* official clock. Indeed, the broad reach of *eBay* invites a host of intercultural communication issues.

Cultural mores, regional values, even differences in geography, weather and climate can set local perceptions, personal assumptions and default understandings about what a descriptor like "excellent" means to different people in different places. Any number of variables can destabilize the deal and substantially change the meaning of things for parties concerned. I'd argue that a widespread reduction in the functional value of personal perspective and judgment is taking place on *eBay*. As with the increasingly superfluous role played by our inner thoughts and feelings, one's immediate, subjective experience and perception no longer applies and, as such, obsolesces early in the game. With this, Meyrowitz' *no sense of place* finds a place.

Like *eBay's* prescription to attend to and adhere to its "official time," when doing business, one also must become attuned to its official geography of nowhere. But this is not a natural inclination. In everyday life, unless one is unusually cool and detached and brutally honest in their assessments, our unique subject positions color and coat just about everything we perceive: ourselves, our friends, our politics, our cars, our kids, etc. It is not uncommon to believe we know what is right and, if our ego is in decent shape, to even think that what we have is good—maybe even the best around.

However, on *eBay*, through some of the mechanisms just described, we are obligated to think more objectively, more universally, and outside of our own experience in a far more systematic way. This leveling or broadening (some have said *democratizing*) effect of electronic media is also a key feature of globalization. It provided the impetus for a substantial part of McLuhan's *global village* idea, through which he figured we would all have access to each other's business. What these *eBay* experiences show us is that this kind of perceptual *outering*—the objective infiltration of one's subjective experience—is coming into its own.

But some important differences remain between the kind of global village now emerging in places like *eBay* and the *McLuhan-esque* version gaining renewed attention in the popular press. McLuhan predicted a smoothing over of media personas represented on screens that have muted detail and generally lack specificity. This is his *cool* media characterization and it requires some qualification in this day and age. I detail the socio-psychological significances related to a renaissance of the written word online in the form of blogging in an earlier chapter devoted to news blogs. But the current status of the word as the primary means of symbolization in a media environment like *eBay* (primarily in the form of lengthy item descriptions) problematizes anew McLuhan's characterization of personalities represented on screens as generally cool and indistinct.

Post-auction feedback details are spatially and socially constrained by the *eBay* interface and emerging social code to limit content to very brief, highly conventionalized accolades. This, in turn, facilitates a generally positive mood for all parties concerned. Pre-auction, conversely, one is prompted to describe items accurately and in great detail from an objective vantage, as well as make clear one's own persona as a stable, rational player in the auction game. This public legibility not only enables but can also predict and, according to the reports of many users, even dictate actions triggered initially through a heightened degree of self-consciousness embedded in the textual format itself. A strange and unprecedented kind of identity stability emerges, and Descartes' hope for a *clear and distinct* subject of analysis comes that much closer to fruition. A "glocal" mode of perception may therefore be the new prescription for success in digital environments. Gabardi (2000) suggests how *glocalization* is marked by:

> [t]he development of diverse, overlapping fields of global-local linkages . . . [creating] a condition of globalized panlocality . . . what anthropologist Arjun Appadurai calls deterritorialized, global spatial "scapes" (ethnoscapes, technoscapes, finanscapes, mediascapes, and ideoscapes) . . . This condition of glocalization . . . represents a shift from a more territorialized learning process bound up with the nation-state society to one more fluid and translocal. Culture has become a much more mobile, human software employed to mix elements from diverse contexts. With cultural forms and practices more separate from geographic, institutional, and ascriptive embeddenness (p. 33–34).

Gabardi's description raises two serious concerns for *eBayers* as the auction site indeed presents its users with a rather profound sort of *catch-22*. First, there is the problem the seller (in particular) has in becoming acquainted with the perceptions and perspectives common to the different locales represented by all players involved. Second is the problem of simultaneously maintaining what is expected by *eBay* to be a consistent, stable self. But these are hardly parallel enterprises. Meyrowitz (2005) elaborates on the problem this glocalizing effect represents for the individual still rooted in one physical locale.

> Although we now often choose a place *more* carefully than people did in the past, our interactions with place have come to resemble what E. Relph calls, in a somewhat different context, "incidental outsideness" and "objective outsideness." That is, the more our sense of self and experience is linked to interactions through media, the more our physical locales become the backdrops for these other experiences rather than our full life space (p. 26).

With the expression "physical backdrop," Meyrowitz conjures the dramaturgical metaphor of Burke and Goffman. What happens to us—our perception of self, and other—when our embodied, socially embedded experience loses its role as the basic context through which we gain a sense of who we are, when it reduces to a kind of staging ground for our actions online? Richard Sennett's (1998) observations that a progressive loss of identity due to itinerant lifestyles and the attendant inability to gain a relatively consistent sense of self might help explain some of this phenomenon. Along with Mead (1934) and the social interactionists, Sennett sees the consistent interaction of persons in face-to-face encounters as the primary way we come to not only know others, but also ourselves (cf. see also Bavelas et al., 1997, and Berger et al., 1973). According to Sennett, something is lost when we no longer have the embodied *other* there to, in a sense, check our own behavior. What is more, the ability to substantially remake the self in a move from one geographical locale (and/or social place) to the next may become a thing of the past given the persistent personal histories maintained in virtual places like *eBay* in the form of detailed item descriptions and always visible user profiles.

These new forms of legibility also enable a new sort of social transparency. But McLuhan's prediction of a kind of electronic *Global ESP* is not what is taking shape. From what I have been able to gather, personal thoughts and intentions are rarely revealed on *eBay*. Legible minds and personalities are not what stream forth. Instead, private thought and intention are functionally obsolesced, rendered quaint side effects of being social. Sociality itself becomes little more than a gambit where it is no longer the thought that counts. To the contrary, through the manifestation of a manic kind of self-consciousness fostered and foisted upon users through the interface, actions become strategic and meaningful inasmuch as they enhance a person's progression toward the practical goal of the purchase or sale, with the prospect of an enhanced feedback rating always in the offing.

On *eBay,* short descriptions and numbers delineating self and other are always, so to speak, in one's face. As Adin's experience wonderfully demonstrates, they stand as a constant reminder—like a personal credit report for the world to see. On *eBay*, others' opinions of us do not fade. An interesting irony and tension therefore emerges between a lagging, stable, modern conception of self, and the fluid, post-modern theory of identity often characterized in the popular press as something endemic to the Internet. Again, as Turkle (1995) opined, "[a] more fluid sense of self [online] allows a greater capacity for acknowledging diversity" (p. 261, brackets added). And Turkle sticks to this assessment in her most recent book. But despite such hopes, the hyper-stable persona may now be surfacing as the pre-eminent being, the successful self in the context of online life.

At bottom, *eBay's* feedback mechanism is a way to quantify and make tangible an otherwise qualitative reputation or personality index accumulated over time and space—one's *credibility quotient*. Traditionally, only a limited portion of our reputation is known to specific others, with even less of it under our own control. The balance of this blind portion of the self usually remains a mystery that is circulated unbeknownst and largely at the discretion (or not) of others who know, or at least know of, us. On *eBay*, on the other hand, so much is revealed. As one very enthusiastic research participant explains below, we see how user commentary on *eBay* often has the feel of empty pandering.

THE HAPPY HEGEMONY: CONTENT AS *CONTENT*

eBay corporate has a vested interest in helping every user maintain positive feedback. Traditionally, the rule in business is that things run smoothly when all players are satisfied. On *eBay,* this rule remains but with even more hollowness. Business runs smoothly when the players simply *appear* content. However, many *eBayers* learn to cultivate a deep interest in the same surface valuations. Answering a short questionnaire that included an item asking participants to describe the process of leaving feedback, *Ned* (a mechanical engineer in his late thirties) was a bit more tongue-in-cheek but no less serious in his underlying concerns than Adin. Freely embellishing the language of the system prompts to make his point, Ned had this to say about the Feedback system:

Have you ever tried to leave neutral or negative feedback? It goes something like this . . .
Click, click, click . . . "I would like to leave neutral feedback, please"
Ebay: "Are you sure you want to do that?"
User: "Um . . . Yes."
Ebay: "Are you really, REALLY SURE?"
User: "YES."
Ebay: "Here, read this, it's important to understand what you are about to commit . . ."

Blah, blah, blah . . .
Ebay: "Now, are you still sure?"
User: "YES, DAMMIT!"
Ebay: "Okay, we'll let you do it, but you need to understand what you're doing is wrong"
Click . . .
Ebay: "Thank you for leaving feedback—it is SO important, blah, blah—have a nice day."
I bet a lot of people just say, "the hell with it!" after maybe the second "Are you sure?" I think this generally elevates the feedback scores of the eBay community and creates a false sense of wellbeing. I mean have you spent any time reading eBay feedback? It's like reading the ramblings of a perpetually cheerful person on Prozac . . . and cocaine!

If Ned's embellishments, and his underlying cynicism, seem extreme, most of the participants in this study seemed weary of the ritual. However, most nonetheless recognized the functional value in keeping up appearances. Hyperbole aside, no news is not good news on *eBay*. Even neutral news is shunned by sellers, and actively discouraged by *eBay* administrators. In *eBay's* official *DOs and DONT's* guide, the final in a list of five items suggests:

> DON'T leave a neutral without contacting a seller first. Most eBayers consider a neutral to be a negative. Regardless, it is a non-positive feedback. Even if the transaction is cancelled, if the seller was courteous and reasonable, then they should get a positive. Neutrals are simply a less harsh form of a negative to most sellers.

As a consequence, feedback on *eBay* is a generally positive (if sometimes half-hearted) history of activities and behaviors tapped out by parties involved. A cursory glance through virtually any feedback profile reveals the nature of this prattle. The bulk of verbal content that makes up the core mechanism for feedback on *eBay* looks and feels superficial. As Ned implies, and as *eBay* itself tries to make clear with its feedback advice, the system generally elevates the feedback scores and pushes the positive. *eBay* feedback is a form of advertising and promotion that gathers around its users in a manner largely beyond their control. Simply put, *eBayers* become the content of their medium.

McLuhan (1951, 1964) made an early move beyond the standard psychological interpretation of advertising, suggesting ads form an environmental surround that contributes to the formation of consciousness and community. "Ideally, advertising aims at the goal of a programmed harmony among all human impulses and aspirations and endeavors . . . When all production and all consumption are brought into a pre-established harmony with all desire and all effort, then advertising will have liquidated itself by its own success" (1964, p.227). To be sure, advertising can begin to work so well with some products and services that so as to render itself obsolete. It is arguable that this has been the case with *Coca-Cola* for decades now—probably longer. Likewise, it seems reasonable to suggest that *MS Windows* no longer needs to be commercially promoted. For all intents and purposes, both are now fixed parts of not only American, but global culture. The brands have become institutions and trans-cultural icons all their own. The names, in short, function eponymously. Coke means soda. Windows means computer/computer system. Google means Internet search. And in terms of capturing (or creating) the essence of Internet commerce, there's *eBay*.

Like small talk at the office, *eBay's* feedback mechanism provides the social lubricant for getting things done. It becomes the important cultural content which, for many like Adin, Andy, Dan, and Tom, is at least as consequential as closing the deal. Andy (the film dealer) recounted numerous sellers who incessantly urged him to leave feedback after the transaction was complete. Andy's crimes on *eBay* are not based on conscious intent. Like Dan, he self-describes as absentminded and easily distracted. But on *eBay* these kinds of relatively innocuous (if sometimes annoying) social infractions are compounded and enhanced in their negative import. Otherwise negligible moments of questionable etiquette are not tolerated there. Offering, perhaps, some moral support for Andy's commentary on this, Adin is convinced that "sellers have to be more polite too—that's standard." Indeed, a hyper-idealized notion of *customer service* comes at us in spades on *eBay*.

Or consider Tom, a musician, electronics buff, and audiophile. Tom's experience suggests that, quite unlike the comment cards made available for customers at many traditional merchants, or the feedback box on the boss's door, the feedback mechanism on *eBay* sets up a situation where personal reputation becomes the supreme commodity. It is a commodity closely linked to *eBay's* reputation wherever longevity in that domain is a serious aspiration. However, as has already been noted, one side effect of this way of being seems to be an increase in a kind of generalized anxiety. There is certainly a great deal of suppression of affect noted by participants. Tom's commentary is perfectly illustrative in this regard. He describes being "forced" to engage in a drawn-out post-auction exchange over digital audio equipment he put up that (to his surprise upon final inspection before shipping) was found to be defective. Tom's annoyance that he "had to be polite to that idiot again" was palpable during the interview.

> I had this portable DAT player that was in great shape. I took a bunch of pics
> and put it up [on *eBay*] with some really technical stock description I grabbed

from another DAT on auction. I added a little testimonial of my own and was really excited because after the bidding started I realized I was probably going to double my money. I almost did, and the guy was really, really happy about this. Then, as I was charging the batteries in final prep for shipping I checked the auto track select function one last time because he asked about this. Apparently this sometimes gives some trouble . . . when the power gets low . . . and the thing was barely able to play for five seconds without stopping in its tracks. I was really pissed, it made no sense, but I still had to apologize to this guy ad nauseam because he really could have screwed me over big time with negative feedback. You can't make any mistakes there.

eBayers work hard and are even willing to pay for this new kind of street credibility. For Adin, the price of a good feedback rating is "forty or fifty dollars . . . I won't argue and fight below that, I'll just pay it . . . it is *very* important." Tracey, a more itinerant *eBayer* than Adin, thought ten or fifteen dollars was her breaking point. Andy, the film buff who makes the majority of his income on *eBay,* figured "[h]ey at this point I've got too much invested as a seller. If someone told me they didn't receive a $100 item and I even had all the records indicating that the delivery was made to their front porch, I probably wouldn't risk it. I'd refund their money."[7] Whether Andy's theory ever cashes out in practice, his impression of himself (and the system within which he operates) is notable.

As Hillis et al. (2006) make clear, "to trade effectively on *eBay* one must seem to possess a stable identity, verified through a user ID and password, a permanent address, and a credit card" (p. 5). But we see now that the operative phrase in Hillis' account is "seem to possess." These features, coupled with considerable and consistently good feedback from peers typically ensures high social standing as a citizen-consumer of *eBay.* "People are more likely to maintain a single *eBay* identity because, as it accumulates positive feedback, it constitutes a key component of their *eBay* brand" (ibid). *eBay* is an interactive environment where identity is written out and accumulates as a record of transactions that is available to all, with the core information immediately visible to the individual member at entry. Through the feedback mechanism, *eBayers* become branded objects, commodities traded and sold by automatic reference, personal suggestion, and innuendo. As such, the individual literally buys into what Roland Barthes described as the *hegemonizing logic of our own commodification* (ibid). Sellers, buyers, anyone who acts on *eBay* become personal brands that are "brand extensions" of the larger site. In that somewhat elitist suburb of *eBay* that I have become familiar with, which sustains a robust alpine snowboarding community, these individuals have names like "bomberman," "borntoride," "trenchdigger1," and "hyperborean." The inhabitants, even some like me who are at best sporadic participants of the subculture, find themselves defined and organized by the categories of goods on offer in that particular enclave. It is a form of human branding—and implicit brand loyalty—within the larger brand-culture of *eBay*.[8]

Andy illustrates this kind of branding well when he suggests how the last four years dealing on the site is analogous to the custody arrangement with his ex-wife:

> Julia can be really difficult to deal with and this directly impacts the time I have with my two kids. It's simple, and I think it holds true even more now than when we were together. If she's happy my life is more pleasant. She drops the kids off on time, she is where she says she'll be and is really just much easier to deal with when my checks get to her without a hitch. But it's sort of a false politeness I'd say.

Andy's reference to a "false politeness" illuminates the forced civility imprinted onto him that seems to be a consequence of being bound up in a relational system based almost exclusively on an unspoken power differential. On *eBay* he is a regular seller of classic comics, 16mm films, and other "old school" media. In his off-line life Andy is a part-time teacher, and single father with visiting rights. In both contexts he describes a requirement to keep others happy in order to ensure that he gets what he wants and needs. On *eBay* this means repeat buyers, off-line this means time with his kids. There is an unmistakable if uneasy détente in both contexts.

THE SINISTER SIDE OF THE GOOD CITIZEN

The degree to which logics of thought and action engendered by the *eBay* environment become adaptive or advantageous to activity inside that domain has been described here in some detail. I'll finally consider some of the ways interaction on the site may actually be disadvantageous to individuals, social groups, and the wider ecology, as patterns of interaction on the site as well as so much of the material by-product of that interaction bleeds into other regions of the on- and off-line world. It would seem that, as templates for civic behavior, the interaction now flourishing on *eBay* could never be sustainable in a wider context that aspires toward anything other than commodity exchange as its core activity. As Meyrowitz (1985) urges, we need to move beyond the standard view that communication media are simply "conveyors of information and begin to think of them as types of social settings or social contexts" (p.93). In short, we need to begin conceiving of any media or media system as formative templates for all kinds of behaviors, whatever the business or the setting. As I was making some final edits to the manuscript for this chapter in my kitchen, I happened to overhear a *Citizens Bank* commercial on the television in the adjacent room. The narrator said that citizenship is a system of social responsibilities. I wondered if that line was the product of Madison Avenue, or whether it might be part of something authored by a notable political figure from history since it made a lot of sense to me when I was thinking about the kind of citizen a place like *eBay* seems to nurture.

Recall *eBayer* Tom's comment: "What? I have to be polite to that idiot again?" This is a reminder of the forced civility now in place—a particularly empty sort of "social lubricant" functioning as operative discursive content. Or, one can take up the devil's advocate position on this and suggest that it doesn't matter what people think any more on *eBay*, a place where subjective drive loses most of its causal role in one's daily life. On this view, personal intention becomes superfluous to practical action. If, from one vantage we can say that *eBay* has created a good and stable society, this assessment loses some validity when we consider both the individual and broader social manifestations that can emerge from these underlying patterns of engagement. Clearly there are similar phenomena now in process off-line as well. These are not isolated social practices.

Impression management, public relations, the wide reach of numerous *dot coms* turned *dot orgs*, and the new cache that comes with virtually any form of social or environmental advocacy has resulted in some very ironic combinations of positive and negative effects around the world. Consider Philip Morris' anti-smoking platform; Walmart's new-found community outreach programs; the Gap's "Red" crusade; Starbucks fair trade component; Honda's *environmentology* campaign; GM's early introduction—and hasty withdrawal—of a successful, consumer-tested electric car; George Lucas' *Edutopia*; the Bill and Melinda Gates Foundation's apparent efforts to accomplish something just short of an end to poverty, hunger, and illiteracy . . . the list grows every day.

On the up-side, we can hope that if self-interest is at the bottom of many of these campaigns, these entities also represent undeniably practical and positive impacts on the social and natural worlds. But the question remains: do these good works simply mitigate more profound and more pernicious activity occuring elsewhere? Such efforts, when viewed from a wider context of whole cultures, economies, and ecosystems may amount to so much lip service. Nonetheless, various kinds of "empty civility" are the latest craze in Hollywood too, with literally hundreds of stars and starlets, media celebrities, and celebrity-qua-opinion leader-citizens dabbling in environmentalism, humanitarian work, and even international politics.

The new reality rails directly against key aspects of the analyses offered in Lillie and Jarrett et al. (2006). It is not so much the exercise of neoliberal choice but something more like Erich Fromm's escape from freedom—leading to the simulation of self in place of some mythic, but nonetheless missing essence. *eBay's* "landscape" is indeed one continuous advertise-ment—with its banners and buyers and sellers all acting in unison to fill out the *substance* of its product-persona. Commenting on this always insistent accretion of emptiness, Baudrillard (1998) seems to have captured something profound about our modern society. He suggested that all of culture becomes an advertisement for itself, constantly reaffirming its own existence and legitimacy. Baudrillard points to something like Carey's (1989) ritual form of communication, and seems to describe well what is going on with *eBay*. With the invisible network effect always in full effect, *eBay* is an environment that makes continued participation a tacit warrant for virtually any prohibi-tion, limitation, restraint, user co-optation, or governmental cooperation its administrators deem fit. And in the face of such massive scale, user boycotts and strikes simply do not work.[9] Ironically, with all of the subterfuge, it is the *eBay* members who take on the burden of work persuading each other that the site and all of its environs are safe, equitable—and, increasingly, indispens-able to their lives.

Loosed from the ethical moorings most intrinsic to socially embedded, embodied, face-to-face interaction (cf. Buber, 1977; Schutz, 1972), a trou-bling pattern emerges. Individuals tend to mislay their subject position, along with most of their personal drive, intuition, and instincts that have been tempered, over time, by the tacit cultural codes the vast majority of us live by. French phenomenologist Merleau-Ponty devoted much of his profes-sional career to the sacred nature of this dynamic, articulating the reciprocal, socially grounded basis of human being-in-the-world:

> Our perspectives merge into each other, and we co-exist through a common world. In the present dialogue, I am freed from myself, for the other person's thoughts are certainly his; they are not of my making, though I do grasp them the moment they come into being, or even anticipate them. And indeed, the objection my interlocutor raises to what I say draws from me thoughts which I had no idea I possessed, making me think too (1945, p.354).

Yet there is more than just awareness of a sort of sacredness in this intimate dynamic he describes. To achieve this oneness, said Merleau-Ponty, we require the element of *interruptibility*. In other words, the medium of live talk, of face-to-face interaction has its own inherent logic that is also deterministic and consequential in certain ways. And this, he claimed, was the essence of communication. But this inherently symbiotic relationship between human beings locked in the symbolic embrace of dialogue is quelled, suppressed, and all but lost online. *eBay* certainly short circuits what Merleau-Ponty describes, as any asynchronous communication mechanism does. Indeed, a new communicative concept is captured succinctly in the *snipe* and the attendant notion that one can and should "stomp their competition on *eBay*."

Sniping is a term that describes various methods of bidding in the last several seconds to win coveted items on auction. It has become an industry unto itself, with books, manuals, websites and specialized software devoted to the sure-fire snipe. The whole enterprise betrays an underlying enmity among and between buyers. I was sniped by humans and bots in my *eBay* years, and I even managed to win a classic snowboard by manually sniping it out from under another bidder at 2 a.m. one summer night. Sure, no one really gets hurt but the activity is vicious pure and simple. *eBay* and the sniping craze has spawned hundreds of books, products, and services with titles, names and catch phrases like: "eBay Hacks," "The Maui CEO: Import from China, Sell on eBay, and Live Wherever You Want," "Stomp Your Competition on eBay," "Powersnipe," "Auction Sentry," and "eBay: Top 100 Simplified Tips & Tricks."

With all of this comes increased difficulty in discriminating between ethical and unethical, or good and bad, rule systems, on the one hand, and advantageous and disadvantageous rule systems on the other. In other words, all becomes a matter of local functionality. However counterintuitive, when the key prescriptions for action flow top-down, a tendency toward consent develops out of sheer necessity. Over time, when prescriptions come from the system (i.e., from everywhere), that is, apparently from nowhere in particular, a tendency toward a kind of passive perfunctory consent seems most natural. It becomes, in short, a way of thinking, a matter of survival—a bit like breathing and respiration. What's more, for the authorities involved, virtually any kind of behavior becomes possible, acceptable, and ultimately justifiable within the bounds of that local system. In numerous totalitarian situations throughout history and around the globe, this meant little more than continued existence for the unfortunate many under that particular sphere of control. On *eBay*, this means continued bidding and buying according to the rules and codes and patterned activities now in place—increasingly activities appended to notions of the good life.

As Hannah Arendt suggested in her reporting of the Adolf Eichmann trial (1963), banal evil is the most frightening form of all for the simple reason that it evades detection and defies prediction. It is, in other words, vague and unpredictable. It cannot be *read,* so to speak. What Arendt makes clear is that assessments of good and badness—right and wrong—can and often do lose meaning when we move from the abstract world of ideas, to the concrete world of action. And the added twist on *eBay* is that we never quite see the actions being made. There is a concreteness that spins out of the exchange process, as something real is certainly manifested in the funds that have changed hands and the receipt of an item won, but we must ask ourselves if we care that the underlying machinations leading up to these concrete moments are often, inherently Machiavellian in both design and intent. Personally, I got the feeling after several years that it might not be that therapeutic to dwell in such a place. One could argue that *eBay* is intentional in nature. It has something *in mind.* And there are a few analogues worth mentioning here.

On September 11, 2001 an opportunity arose for the Bush administration to greatly expand its agenda at home and abroad. The bet was that Americans would relinquish some concern over freedom of action, privacy, and civil rights in exchange for security and stability. Surveillance and control in the name of security and safety—that was the bet made in Washington. And the wager delivered in spades. With the establishment of the *Department of Homeland Security* as well as widespread alterations to municipal (especially mass-transit) security rules and regulations, a popular acceptance grew around the notion that *if you aren't doing anything wrong then you have nothing to be concerned about.* An old trope found new life as localized pockets of resistance to the new initiatives quelled in the midst of a nationalistic fervor reminiscent of the kind of insidious social cohesion that was able to be sustained from the beginning of WWII through the McCarthy era. And like much of the operative communication on *eBay*, the submission of *hearsay* becomes a valid way to proceed with the *Military Commission's Act* passing late one night in October of 2006 to a government asleep at the wheel and a public largely unaware of the implications. With the signing of the Foreign Intelligence Surveillance Act in July of 2008, which includes retroactive immunity for any telecommunications companies that helped the U.S. spy on Americans in suspected terrorism cases, the task (the intentional structure of the system) seems almost complete. In all three cases, we see the formal institutionalization of what essentially amounts to gossip and innuendo as a means of censoring individuals, determining the right or wrongness of specific behaviors, and ultimately enforcing corrective action.[10] This is precisely the modus operandi at the core of *eBay's* ethics.

PROGRAMMING A PLEASANT PANOPTICISM

Through the modeling of good civic behavior to facilitate the online auction and shopping experience, *eBay* functions very efficiently as a social control mechanism. But we must remember that "[e]verything that human beings are doing to make it easier to operate computer networks is at the same time, but for different reasons, making it easier for computer networks to operate human beings" (Dyson, 1996; 10). Echoing many of the caveats Norbert Wiener lodges early on in his seminal *Cybernetics* (with the oft-forgotten but sobering subtitle: *or control and communication in the Animal and the Machine*) Dyson offers a distinctly media ecological argument, pointing out that "[s]ymbiosis operates by way of positive rewards. The benefits of telecommunications are so attractive that we are eager to share our world with these machines" (ibid). Systems like *eBay* indeed facilitate the control of humans and encourage a willingness to submit to that control. William Burroughs' policeman is inside and outside, and everywhere in between . . . and slowly but surely, we come to believe that it's *OK*. On the street and on *eBay* we may find that public interaction reduces our roles to that of fellow supervisors. Any social fabric, whatever its primary medium, eventually tears itself apart when based on fear (of detection, of reprisal, etc.). History offers us substantial evidence that such platforms are unsustainable. Commenting primarily on the social pathologies intrinsic to Soviet Russia, Bertrand Russell noted how "either a man nor a crowd nor a nation can be trusted to act humanely, or to think sanely, under the influence of a great fear" (Russell, 1992; 98). If Russell could only see us now.

Due to intrinsic design features of certain media, predictions can be made concerning their social and psychological effects or side effects. It has been argued well (Eisenstein, 1977; Ong, 1973) how the shift to mass produced printed material at the end of the fifteenth century prompted a shift toward abstract thought, objectification and even nationalism. I've argued how *eBay's* formal design features (features representative of many new digital communication platforms) may systematically facilitate a social reversion. Virtual places and spaces like *eBay* can retrieve by reversing or flipping back into moments from our social past that are, with the benefit of hindsight, now plainly seen as non-functional, maladaptive, and ultimately non-sustainable forms of life. Nonetheless, we seem to be moving ahead full-steam, back to some warped 1950s Americana, when the appearance and talk about a golden age, of being a good neighbor merely sustained the appearance of a peaceful, healthy, and orderly society. Such was the mythology of the time. However, simmering just beneath that surface also loomed a host of suppressions, repressions, and social anxieties. And the evidence is mounting which suggests that modern mass media have exacerbated trends in these forms of experience (Donaldson, 1997; Showalter, 1997; Schor, 1991; Oldenburg, 1999). Now, with the always-accessible-always-on nature of communication made possible by the Internet, the ability for social pathologies to proliferate seems to be significantly enhanced.

As *eBay* renders specific behaviors normative, academic questions concerning the interplay between sociogenic and psychogenic forces become moot. These questions now have a quick and clear answer. In the midst of hundreds of millions of users acting in kind, an elaborate, emergent rule system gains initial traction, progressive acceptance, and then legitimizes itself along the way. Given this, we can no longer seriously entertain the notion of individual psychological patterns setting off a cascade of social effects to impact larger collectives. No *one* can make anything happen on *eBay*. Instead, an amorphous new kind of collective intelligence sets the pace, design, and changing structure for what, in an earlier time, used to be considered individual psychological types. Now a top-down social organism with, ironically, no central node or localized command center to speak of fixes hitherto human subjects like Gutenberg's press stamped out its primordial font types in progressively widening patterns of dissemination.

I think these processes conjure more than just a naturalistic metaphor. They are unmistakably ecological, systemic, and ultimately biological in their emergent morphological patternings. Since the dawn of global sea trade Zebra mussels and blue-green algae have continued to infiltrate our coastal regions, lakes, and river beds, choking out entire strata of both plant and animal life. Might *eBay*, or an *eBay*-like social ethic, adapted originally to the logics and functions of its original domain eventually colonize the shores of our social and political lives making it parasitic upon the idea and the actual doing of democracy? With the progressive transmutation of our symbolic space from the spoken to the written to the visio-iconic, our new thing-ing abilities allow the representation of human beings in a very different light; at a distance, different, objectifiable, manipulable. Likewise, in the transmutation from atoms to electrons, the nature of being human has shifted even further from *a way of being* to being *a kind of thing*. Heidegger would surely interpret *eBay* as a massive framework (*gestell*) that sets human beings up as its raw material (*bestand*). Or as Burroughs once opined: "*Human . . . is an adjective and its use as a noun is in itself regrettable*" (in Ballard, 1997; 135).

CONCLUSION: THE SPECTER OF UNSUSTAINABILITY

Several aspects of *eBay's* progressive flooding of the Internet have not been addressed here which may hold profound ramifications for our lives off-line. While the on/off parlance can easily be misused so as to reify a false dichotomy, it might be useful to think in these terms for a little while longer. For one, we are seeing a vast, popular movement toward collecting as a pastime, or fervent hobby, even a way of life on *eBay* (Hillis et al., 2006). And so while not at all new, nor a typically problematic phenomenon in and of itself, it is arguable that the habit and felt-need to accumulate is becoming a kind of pathology for many *eBayers*. An obsessive-compulsive tendency was noted by more than half of my research participants, and I must admit that twelve snowboards (five of the alpine variety alone) also borders on an "unhealthy" compulsion. As such, we might be witnessing the emergence of an explicitly materialist form of OCD as an essential part of the *eBayer* identity, and a core value of the *eBay* brand. When the virtual and real worlds open up to us as our personal marketplace, it is only natural that every want turns into a need, and every need holds the potential of becoming an obsession. Negroponte's *Daily Me* concept takes on a dark hue as we can never seem to pull enough toward ourselves. And even if an obsession is, by definition, insatiable, there are no worries there since *eBay* can always provide.

Most people will admit that a very satisfying, almost visceral experience often accompanies purchasing something online or off—whether a new piece of jewelry, clothing, or a gadget not yet possessed. Bringing the object home (or having it delivered) typically results in a temporary catharsis and a general feeling that one is doing something, even that something important has been accomplished (Perry and Massie, 2007). And now this sense of agency, of doing and manipulating, of getting something done has been streamlined and rationalized. Recalling Carey's (1989) suggestion that machines have "teleological insight," we have to take seriously this particular *question concerning technology* (cf. Heidegger, 1977). Indeed, *what* do our devices and systems seem best suited for? What are they good at? As the superlative manifestation of computer technology, the Internet may have realized one important part of the essence of computation. As more professional sellers adopt the online auction model, future investigations may well reveal that computer technology has found one significant part of its telos in *eBay*: the logical fruition of the science of selling, with its core values of calculation, efficiency, prediction, and control. It is, perhaps, the lack of questioning of these ends that Postman laments at the conclusion of *Amusing Ourselves to Death*:

> Although I believe the computer to be a vastly overrated technology, I mention it here because, clearly, Americans have accorded it their customary mindless inattention; which means they will use it as they are told, without a whimper. Thus, a central thesis of computer technology—that the principal difficulty we have in solving problems stems from insufficient data—will go unexamined. Until, years from now, when it will be noticed that the massive collection and speed-of-light retrieval of data have been of great value to large-scale organizations but have solved very little of importance to most people and have created at least as many problems for them as they have solved (1985, p 161).

Postman's cynical prognostication may be coming to pass—especially where information-centric organizations like *eBay* and other similar digital manifestations are concerned. But can such patterns and relations be avoided? Can we turn our energies to other kinds of productive behavior? Or, is meaningful social interaction online a phantasm, an empty dream of the happy modernist? I have argued here that it may be structurally impossible to sustain the kind of dialogic necessary for substantive social interaction that might eventually translate into positive, practical, real-world outcomes. The structural incompatibilities between the way humans have evolved as communicative beings and the way the Internet has, in a sense, evolved *out of* certain design features basic to computational architecture may often preclude genuinely human-qua-humane interaction. Such interaction seems to require embeddedness in real-time face-to-face exchange (by no means a sufficient condition for civility, though perhaps a necessary one). The electronic commodity transaction rooted in the typographic mode, on the other hand, seems ideally formatted for objectification. Buying and selling might just work better this way—largely unseen, mostly unknown, and at a distance.[11]

At its simplest then, *eBay* is a tool that enables the exchange of goods and capital. But as McLuhan made clear throughout his writings, we shape and fashion our tools and machines and they in turn shape us. After a year and a half thinking about these issues, and participating, lurking, and talking with people who frequent *eBay*, I get the sense that we can build a sustainable community centered around the exchange of goods and services there. But it is not clear to me what more we can build from this. *eBay* is probably best understood therefore as a panoptic, socio-technical commodity exchange system.[12]

Two collective values underlying life and work in the United States at the beginning of the twentieth century through World War II were frugality and thrift. This ethos was then inverted and many of the associated activities were abandoned by mid-century, as consumption assumed its status as the pre-eminent post-war virtue. By century's end, consumption becomes one of the most popular pastimes noted in national surveys and, for many suburbanites, a reliable means of ebbing ennui (Desmond, 2003). Rampant, conspicuous consumption remains a core part of the American ethos and identity. *eBay* comes on the scene in the mid-1990s welcoming a population of conspicuous consumers.

Today, as *eBay's* membership moves beyond the three hundred million mark there is no limit to what is on offer, and subsequently, no end to want and need. This is a new breadth and scale of collecting orders of magnitude beyond anything we have seen before. Anyone can collect anything quickly and successfully if they are only willing to put in the requisite time, effort, and funds. And yet good deals on *eBay* are increasingly difficult to find as the online auction context precludes any debate about fair valuation or market price. To be sure, on *eBay*, things are not always what they seem. As distributors and even mass producers colonize the online marketplace, value is no longer connected to the law of scarcity. If this is the new free market, it is a skewed form of freedom for a select few.

Finally, there is one last and perhaps even more problematic aspect to commerce and communication on the site that has not been addressed. Like any commercial purchase, payments made through *PayPal* or the more traditional methods of payment like cashier's check or money orders effectively conceal the material processes that accompany these transactions. Only a very small percentage of the deals made on *eBay* concern items that remain in digital/ electronic form that do not require, for example, any notable release of carbon into the atmosphere (e.g., some music files, electronic sporting, music and event tickets, travel and service vouchers, etc.). Most of the material on offer is indeed material—that is, composed of atoms—and therefore must be packaged and transported in some fashion. The environmental impacts related to a massive increase in the number of individual parcels shipped to separate destinations have yet to be seriously discussed. What used to be a purchase along with many others at a local shopping center, warehouse, or garage sale has now fragmented into millions of discrete transactions. What, for example, would *eBay's* carbon footprint look like? Is it possible to gauge the environmental impact of a mostly virtual corporation? An answer there will undoubtedly require some sophisticated metrics. Likewise, consider the fact that apples trucked from New York to Boston can and often do have a greater negative impact on the biosphere than the same variety shipped overseas from New Zealand. This all suggests that the underlying realities are often counterintuitive to say the least. Indeed, nothing is obvious here.

The wider environmental effects associated with the collection, packaging, transportation, and eventual disposal of items, and all additional materials utilized in the transactions that take place between individual sellers and buyers on *eBay*, will need to be figured into the larger equation. It seems questionable that these are sustainable activities. To the contrary, the piecemeal way *eBay* items move around the country and globe likely results in substantial detrimental effects to the natural environment. This is just one more reason why *eBay* will probably never be a reliable model of community. But again, these problems may run in tandem with other underlying socio-environmental side effects of digital commerce and Digination in general. Similar to the manner in which the entire biosphere is transmuted into *standing reserve*, a disembodied textual mode of interaction seems to short circuit the *We*-relationship, the *Thou* situation of Schutz and Buber, and that sacred co-presence described by Merleau-Ponty. On *eBay*, all is reduced to an otherness that remains at a distance, abstracted, distilled, and easily manipulable. Joseph Stiglitz, chief economist at the World Bank from 1996 until 1999, recently turned his eye to the hidden costs of the War in Iraq (Stiglitz, 2003; Gross, 2008). Like Stiglitz's recent inquiries, there is an argument to be made regarding the hidden costs of *eBay*—costs related to the health of the cognitive, emotional, social, political, and natural worlds we inhabit.

NOTES

1. In 2008 *eBay's* total market capitalization was about $40 billion according to Newsweek tech-business writer Daniel Gross (www.newsweek.com/id/138221).

2. For being so quantified (number of sales, feedback ratings and quotients), *eBay's* rating system only provides three options: positive, negative, and neutral. In the end this is a pretty unsophisticated meter. It is akin, as Adin opined, to saying: "I'm more happy than unhappy, therefore I must be happy." For many Americans this was the quandary they found themselves in when the president asked, in 2004, if we were better off at that point (rather than four years prior). A systematic oversimplification and distillation of personal affect results. However, not all online commodity-based communities are the same in this regard. For instance, *Amazon's* marketplace offers a 1–5 rating system. If still certainly constraining, this is slightly better in terms of representing personal impression. As of May 2008 eBay no longer allowed its users to post negative or even neutral feedback. This development is right in keeping with the underlying logic of the system.

3. The terms "modern" and "stable" in no way imply the existence of an authentic self or core identity elsewhere. Rather, they merely point to the emergence of a highly functional self, adapted to the specific requirements of the *eBay* interface within which users operate—itself a highly rationalized, stable, consistent environment.

4. Certainly, many of the safest, most serene looking social spaces of the world hide a rigorous police/paramilitary structure just below the surface (consider Zurich, Singapore, the mock-up of Venice in Las Vegas, even the *River Walk* in San Antonio Texas). Do nothing wrong and there is no trouble. Make trouble, and count the seconds to apprehension and potential processing.

5. Foucault (1973, 1976) describes the persistent information structures that accreted around private individuals in the eighteenth and nineteenth centuries as mechanical printing became commonplace and the medical establishment became an entrenched institution part and parcel to Western culture. Today, elaborate data edifices are preceding, forming around, and following us everywhere we go. Consider the doctor's office, the transit terminal, the shopping mall, the classroom, and office space. Telephone companies trade data with mortgage lenders, who sell or trade or "lend" our information to potential employers and credit card companies and law enforcement agencies. *Google Analytics* is a free, downloadable tool that allows a staggering level of detail concerning Internet users' browsing and click behavior. Teachers can track and even predict their student's behavior via online

course management systems at the level of the LAN and now potential employers, guidance counselors, and law enforcement officials are mining social networking sites like *MySpace* and *Facebook* to see what might be revealed at the level of the WWW. Certain security enhancements are one obvious benefit motivating these trends. Indeed, medical records and credit reports have alleviated some of the risk and loss sustained by lending institutions and insurance companies, but will this kind of top-down control make for good citizens off-line as well? The establishment of a singular, stable self has always been more the stuff of fiction—at least until recently. What is so new and different about *eBay* is the explicit, quasi-public legibility of the personalized data sets it houses.

6. For example, there was a palpable, felt-need to do the right thing after I sold a snowboard on *eBay* for more than $300. But my urge to do right ran beyond just being a good seller/deliverer to my buyer. It was also spurred by an awareness that I would surely return at some point, to list, or to buy, and that my user profile would precede me with every click. While functionally achieving very similar ends for the larger interests concerned, the most significant difference for the consumer-citizen between traditional consumer profiles and *eBay*-type profiling is that the former allow very limited access to the larger group.

7. Whether these personal theories actually cash-out in practice is secondary to the mindset the environment engenders. *eBay* subtly prompts this kind of subservience to the system.

8. On *eBay,* self-reflection does not follow the typical pattern that most digital culture commentators still often espouse. It is not "I don't know who I am" (the existential angst that typifies the post-modern condition, where a vertiginous self-consciousness is on high), or "I am who I am" or, conversely "I am who a localized collection of significant others say I am." In all three cases the self-awareness or self-consciousness component is present, but it is largely idle or out of the immediate sphere of influence of the agent in question. Rather, on *eBay* the configuration becomes "I am who I have to be in order to continue operating in this domain" or even, "I am as *it* or *they* (all other *eBayers* combined—i.e., *eBay* itself) wish me to be."

9. There have been a few ripples on the water. The boycott in the late 1990's brought on by changes in auction policy (including the establishment of the bid "reserve" amount and *"Buy It Now"* option) resulted in only a momentary lull in traffic on the site. *eBay* also had some trouble sustaining its communitarian ethos after it absorbed *PayPal*, the electronic payment service in 2003. However, the seemingly endless stream of personal and collective critique found on blogs, homepages and *YouTube*, or the more official protest sites like *paypalwarning.com* have not resulted in any appreciable dent in *eBay's* business. The network effect is just too robust—and membership still grows steadily at the time of this writing.

10. Does any set of behaviors considered dangerous or even unappealing ever remain fixed? The answer is clearly no. Yet, the absurd contention that it does was perhaps codified most succinctly in *An End to Evil*, a book written by Bush advisors Richard Perle and David Frum in 2003. Unfortunately, there is no end to evil when those in a position to define the term find advantage in broadening its application and scope when it is deemed advantageous.

11. *Craigslist*, the free community-based omni-transaction site, uses a very different model of online commodity exchange. If the initial connection is made in the very austere textual format the free-access site maintains, the dicker/haggle remains in both subsequent e-mail exchanges leading up to an agreement, and perhaps during the ultimate face-to-face meeting that results before final sale. A similar catharsis often occurs with the attainment of a new possession, *Craigslist* simply hosts a much messier environment that is also less secure. And compared to *eBay*, it is a refreshingly tactile experience overall.

12. An *eBay* ethic, that is to say, an *eBay* citizen's duties are very much like what is now being called for by some public leaders. It is akin to the plan by Texas Governor Rick Perry to install night vision cameras along his state's border with Mexico. A prototype surveillance station is already up and running. Brought to fruition, the initiative would allow public access and control to the remote cameras via streaming web connection, inviting anyone interested in spotting illegal aliens attempting to make their way across the frontier a chance to participate personally and in real-time in the national security apparatus by reporting infiltrations to a toll-free number (Metro, 2007; p. 8). In August of 2007 I was allowed to demo the *Black Wolf* remote infrared surveillance system produced by EMX industries, a small defense contractor bidding on the Texas border plan. In Boston, I controlled a weatherproof camera perched atop a mobile platform in Texas, and followed a small dog snooping around along the Mexico side of the river bank at one in the morning local time. It was, put simply, a surreal experience.

Chapter 11

Media Ecology and a Biological Approach to Understanding our Digination

"Everything that human beings are doing to make it easier to operate computer networks is at the same time, but for different reasons, making it easier for computer networks to operate human beings."

—George Dyson, *Darwin Among the Machines* (1997)

IDENTITY, ORGANIZATION, AND PUBLIC LIFE

Eight of the preceding chapters were each devoted to a specific technological artifact or application. I considered some of the differences and similarities between these technologies by framing them along the three themes that make up the book's subtitle. We are moving headlong into a world constituted more everyday by digital information exchange. Given this, we need to be cognizant of the changing nature of identity, the different forms of social, cognitive, and political organization now emerging, and what it means to be a citizen of Digination today.

Consider the question of human identity. As the e-mail and blog analyses reveal, there does not seem to be any essential way of being that exists, or is even possible to describe, outside of some context of interaction. This is in keeping with most post-modernist formulations—humans define themselves, or are defined by others, in relation to some wider situation or set of circumstances. Some of these situations and circumstances include the social and physical contexts we find ourselves in every day. But contexts can also be symbolic. They are often also historical, ideological, political, or media environments (or any combinations of these).

For instance, it was shown how my Mohawk research participants made sense of themselves as e-mail users through a cultural lens. But this cultural lens was in part constituted by a technology that seems, by most of their personal accounts, antithetical to things Mohawk. While there was evidence of some technological demand characteristics and side effects rooted in and emanating from certain features of their e-mail system (most notably, the subject field and composition window), it was their awareness of themselves as Mohawk (traditional Mohawk) technology users that determined their experience. So if a kind of technological determinism was happening there, it was a very subtle form. Without lapsing into tautology, I can argue that it was the reflexive nature of their technology use that determined their technology use. In other words, certain formal features of the medium helped them orient to, and reify, "formal features" in themselves. I interpret this as a *soft* or *weak* kind of technological determinism and suggest such an analysis can do much to illuminate some of the ways both group and personal identity conceptualization operates through (and because of) media environments. In this case, it happened to be a small population of new community builders, but its likely that such effects are generalizable to other populations in different social and media contexts. I address just a small collection of these in this book. They are contexts with a host of different technological biases and logics including e-mail, blogs, portable digital music devices, podcasts, Internet news sites, search engines, classrooms, laptops, smartphones, Twitter, and *eBay*.

As Mead (1934) implied, if there is any sort of *nature* to human identity or to being human, that nature has always depended on the relationships we maintain between self and society. Today those relationships are increasingly mediated by digital communication technologies of various sorts. Given this, we begin to see how new ratios and forms of intentionality and action are in the offing. Who and what has intention is a big part of what Digination, as a cultural process, stands to change. What are the dictates of the human? What are the dictates of the machine? Communication, sociology, anthropology, psychology, and philosophy all hold interest in these kinds of identity issues or "boundary questions." Where my Mohawk friends' confrontation with an early e-mail system suggests that some groups can retain more autonomy in the identity-building process, most of the subsequent chapters indicate, at other times and in other places, that we can also be subject in much more deterministic ways to the formative properties of a technology.

So if these analyses make one thing clear, it is that we are always in the process of *becoming*. They suggest that the human animal is not finished (in terms of our evolution), nor are we complete (in terms of being whole, discrete beings). As the podcasting, Web search and *eBay* chapters tell in particular, our sense of self becomes more ephemeral and therefore seems to lose causal force as we immerse ourselves deeper and deeper into the digital interface.

I write with a sense of urgency in suggesting that we must not only be aware of these environments, but also remain vigilant regarding the nature of their effects. We need to understand our Digination if we plan to have some say concerning what it is we can say and do as individuals and, more importantly, where we're headed collectively—*as a species.* I emphasize this point because there are some new species roaming the world, and they are itchin' to breed.

The second theme threaded through the book concerns the way digital technologies and the contexts they represent actually constitute organizing principles. They reorganize the way we perceive, the way we think, and the way interaction unfolds. We looked at how sending and receiving e-mail messages, reading blogs, and listening to music and voice on the move can change the way we not only think about ourselves but others too. We saw how one's awareness of fellow buyers and sellers in the online auction game is framed and formatted in fairly proscriptive ways by the user interface. And this isn't to say that some group of nefarious interface designers put specific "intentions" into their construct. Instead, what we discover is that much of what is occurring on *eBay* is an accident, an emergent by-product or side effect that spins out of a user interface that is always a theory of how people will behave.

To consider one analogous example, recall the first SONY Walkman models. Many were designed with two headphone jacks. But then things changed. The device morphed into something else because when mixed with real people who have their own ideas (and so often don't do as suggested, instructed, or told) what tends to emerge is a tool used in an unexpected way. In other words, a new tool! In this case most Walkmans were used for solo listening. And so the meaning of the device changed in the midst of its use, with subsequent models being redesigned with a single phone jack. The reader will also recall a similar reciprocal effect in the Mohawk e-mail example, and the history of the answering machine is another telling illustration of the meaning-as-use axiom.

E-mail, text message, and image files we can access on a smart phone or PDA each organize our thoughts and help us store, retrieve, and make sense of memories in unique ways. They change the way we think about ourselves and others, and, in so doing, these technologies are also altered in the process. How many teenagers do you really know who use their phone as a *phone*? These stories are about the reorganization, the blending, and bending relationships between self, other, and artifact.

The Internet news and search engine chapters also describe how information is explicitly organized *for* us. These new ways of organizing information constitute templates for action. That is, they often function as prompts for different modes of thought. Consider how a linear and sequential habit of mind is exercised on a text-based medium like e-mail. Compare that to the different kind of thinking that spins out of regular consumption of an image-based website, or the online version of CNN's headline news, or a blog-turned-vlog where non-linearity is the name of the game, the "rules" of cause and effect are largely suspended, and knee-jerk, emotional responses typically hold sway.

I've demonstrated how all of these different media forms help us organize how we think about and act in the world (in both positive and problematic ways), and how we orient ourselves to persons known and unknown along the way. In these respects, all of the media and mediating technologies discussed not only represent different ways of thinking and being, but they are also consequential to what it means to be a social actor, a political agent, a parent, teacher, student, business associate, or friend. Media are organizational because they modify perception, thought, memory, physical spaces, and human identities. They alter definitions, and recast what it means to be a member of a public, or publics, and offer tacit advice on how to be a good and productive denizen of Digination.

THE HUMAN-TECHNOLOGY RELATION
AS A LIVING SYSTEM

It is the ecologically-minded observer who remains aware of the ongoing give-and-take between media and people. Let's consider what it could mean to say that there is something systemic, something even biological about the way we interact with technology. This should lay any lingering confusion over the meaning of technological determinism to rest. To begin, we return again to a seminal work in the media ecology tradition.

The notion of reciprocal relation is implied in much of Gumpert and Cathcart's *Media grammars, generations, and media gaps* (1985). The authors suggest that a medium's formal structure results in distinct "grammars," "codes," and "conventions." These terms deal with certain constraints bound up in a technology—the ways it must be used in order to be used *properly*.

Gumpert and Cathcart contend that: (1) there is a set of codes and conventions integral to each medium; (2) such codes and conventions constitute part of our media consciousness; (3) the information processing made possible through these various grammars influence our perceptions and values; and (4) the order of acquisition of media literacy produces a particular world perspective which relates and separates persons accordingly (1985, p. 24). In other words, patterns of regular and sustained media use, as opposed to biological age, ethnic origin, or nationality, might even be the best means of understanding the similarities (and differences) between many otherwise ostensibly disparate populations today.

Gumpert and Cathcart's last item, the idea that the order of acquisition relates and separates persons accordingly, is particularly useful in helping us understand our Digination. First, and for quite practical purposes, we should think of media as cultures in the sociological sense implied by the authors. Second, we should also be thinking in the biological sense of cultures as containers that support the growth and propagation of particular kinds of organisms.

Inspired by the many instances of breakdown and loss endured by technophobes and technophiles alike, and borrowing terminology from biology and other natural sciences, future media research should consider some of the interactions we maintain with our machines along the lines of sensory balance (McLuhan, 1964; Ong, 1982). So many people today engage with media that represent heavy biases toward images, icons, and other pictographic information. This can be problematic in the ways I've described in Chapter 6 so I will refrain from recounting those details here. The point is that we might now be in a position to draw out a typology of relations (i.e., symbiotic, parasitic, sustainable, pathological, etc.) between human beings and the technologies they employ in the process of *getting things done*.

Extending Gumpert and Cathcart's notion of media generations, we learn that different media use fosters different ways of representing and experiencing the world as we go about getting things done. Taken to fruition, Gumpert and Cathcart's argument suggests that behaviors stemming from immersion in, or communion with, various media can become fixed over time as a priori constructs and repertoires that are cognitively "stored" among consistent users of those media, forming a cultural community or *media generation*. If deemed useful, those constructs and repertoires are propagated (both by explicit instruction and more tacit, perfunctory mimicry) to subsequent members of a population, thereby spawning the emergence of new kinds of people and media (i.e., new organisms), and the cycle continues. The fax machine, telephone answering machine, and early versions of the Sony Walkman exemplify the reciprocal, formative relations between people and their machines, and the unfolding realities of these relations further bolster the appropriateness of a biological approach to media studies. The Walkman story was touched on above, but let's consider these examples a bit more here to clarify this notion of a priori constructs and repertoires.

The successful patenting and then commercial failure of the fax machine in the mid-nineteenth century hinged upon a social and institutional *need for speed* (of information transfer) that did not yet exist. Means of transporting people and the general pace of life at the time did not prompt many folks to conceive of near-instantaneous document transmission as any real virtue. It took nearly a century, with the eventual ability to move bodies at much greater speed and distance, for the fax machine to become a meaningful invention, let alone a useful communication tool.

Likewise, while a device called a telegraphone was invented in 1898 as the first practical apparatus for magnetic sound recording and reproduction, it wasn't until William Müller patented the first automatic answering machine in 1935 that the popular need for such a thing began to take shape. Even then, the device's largest group of users seems to have been Orthodox Jews who were forbidden to answer the phone on the Sabbath. The widespread marketing of answering machines beginning in the late 1960s marked the introduction of a device designed to replace the human answering service and allow people greater convenience and a new kind of control that accompanies the enhanced organization of incoming calls while away from home (or not).

Indeed, it did not take long for the mutual transmutations of the answering machine and human user to begin. If someone was home while a recording was being made it soon became clear that one could simply "screen" an incoming call (and caller) during or after the message had been successfully recorded. Call-screening simultaneously altered the design function of the machine and prompted a reciprocal change in its users. Perhaps the most significant change was a small perceptual shift that had far reaching consequences. One feature of this shift was a heightened degree of self-consciousness regarding assessments of our own social positions, as well as the personal assessments we make of others. Caller ID boxes emerged later and the cyclical changes continued with caller ID technology embedded into home-based hand-helds and cell phones. Similar reciprocal synergies went into play when the first series of Sony Walkmans hit the U.S. market in the late 1970s. As pointed out above, many of the early models incorporated two headphone jacks under the assumption that listeners would often come in pairs. It was later discovered that consumers were more interested in the option of enveloping themselves in private soundscapes so the second jack was seen as redundant and removed (Hosokawa, 1984; Du Gay and Hall, 1997).

Since an entire chapter in this book is devoted to a phenomenological analysis of everyday uses of portable digital music devices, it should suffice here to say that the recent history of this particular genre of communication technology reveals that pattern of design-use/alteration-redesign over and over again. Reciprocal change in human-machine relationships is another central feature of Digination.

Of course, not all of these constructs and repertoires, and the rules for action often built into the design of a technology are always rigidly determinative of behavior. We see this clearly in the history of the Walkman. Instead, technologies offer us cues for living and serve as resources, principles, and grounds for thought and action that cultural members employ when confronting situations. Scheflen (1979) referred to these principles and grounds as "programs," and suggested that while actual occasions of interaction may veer from these cultural routines (via those constructs and repertoires), some set of cultural routiniza- tion nonetheless forms the basis for interactants' expectations for behavior in communication events. While mostly referring to non-technological situations, Scheflen's conceptualization nonetheless helps us conceive of communication media and all the attendant apparatuses (e.g., phones, cameras, microphones, audio devices, search engines, etc.) as different "species" we interact with since they are designed also to find affordances or meaning, to exchange matter and/ or energy (i.e., particular patterns of organization) in their environments.

Now while this might sound mysterious, there is in fact nothing occult going on here whatsoever. It simply refers to the meanings of stimuli that are registered by (i.e., important to) the particular organizational structure of the perceiving thing or entity. Consumer-grade digital cameras contain a charge-coupled device (CCD) that detects photo emissions along a segment of the electromagnetic spectrum that makes their outputs perceivable by the human eye. Infrared cameras register along another. Traditional (analog) radio tuners pick up frequencies along the RF range of that same spectrum. But the combination of analog transmitters and tuners led to overcrowding over time. Digital radio transmitters and tuners allow much more finely delin- eated access to the spectrum and effectively creates infinite space for content providers. Search engines register the hubs and authorities of the Web that their heuristic algorithms compute.

I'm an English speaker who gets along in French and knows some Spanish too, but who does not comprehend Arabic (or Swahili, or Chinese, or a thousand other languages for that matter). By virtue of my language use alone I am a cyborg. Each of these technological systems bootstraps our cognitive abilities and extends our senses. They help us think and feel. And so I'll end this section with a complicated little compound question: *how are you thinking and feeling today?*

THE INTIMATE RELATIONSHIP BETWEEN TECHNOLOGY AND THE NATURAL WORLD

Understanding our Digination along these lines can provide valuable insight into what is currently happening in our media environments, and how that might have something to do with what is happening in and to our natural environment. I address this concern briefly at the end of Chapter 10, but we should draw out some additional significance in this relationship.

Recall the traditional notion of communication and transportation going hand-in-hand in allowing a kind of bringing-together, communion, and feeding of the world. The idea is intuitive enough. If one assumes a bird's-eye view of the world they would see something that appears decidedly organic, dendritic in its layout and tendency to radiate outward and propagate from one region to another. Ironically perhaps, but especially with infrared-enhanced photography, the lakes, rivers, and streams of the interior of the United States is reminiscent of a vascular network in a hand, or foot, or leaf of a tree. The image would reveal the Mississippi River forming the central vein. The *Google Maps* and *Google Earth* mapping options with a variety of "overlays" allow renderings of most of the planet's surface and actually make this kind of perspective simple to achieve. In the chapter devoted to search engines and the last chapter concerning *eBay* I suggest that this notion of *mapping* as a kind of top-down element of measurement and control has become one of the primary control mechanisms of Digination.

If we think about the nation's waterways we see that these "channels of life" were for some time synonymous with channels of communication. Supplying the required hydration, caloric, protein, and information content to all manner of biological and technological system, the heyday of the waterway in this capacity straddles the nineteenth and twentieth centuries. Prodded by the rains, our lakes, rivers, and streams worked together in the nourishing of the country. And this is about when the wires began, in turn, to run along the waterways and roads, paralleling, and then extending the much slower acting physical nervous system with a faster, high-twitch electronic replacement. This analog electrical infrastructure is in turn replaced with a digital system toward the end of the twentieth century.

The history of the Hudson River in New York State provides a good example to illustrate the evolution and technological extension of a natural, organic communication system. In sequence came the roads, then telegraph, then telephone, and electrical lines. Also, the Erie (East/West) and Champlain (North/South) canal systems are appendages or extended prostheses of the Hudson that were eventually enhanced with electrical lines and, much more recently, fiber-optic conduits and cell towers. Along much of the Hudson today, the nineteenth century rail lines are being widened and integrated into State recreational facilities and parks, with the original rail beds converting nicely into pedestrian and bicycle paths. Similar projects have been completed all over the country. The latest alterations to these facilities are wireless repeater antennae installed along the ridges and tree tops banking the rivers. And the keen observer will sometimes spot a few of those quasi-cedar-looking artificial trees appearing just a bit out-of-place along these routes and interstate highways. These are just more efforts on the part of infrastructure designers to blend our digital prostheses into the natural world.

In other parts of the world the latest manifestations of high-tech/high-speed transportation are even running further afield of their natural conduits. In early July of 2006, the Chinese government opened the world's highest railway (peaking at around 12,000 feet) linking Tibet with the other parts of the mainland. This new bullet train is a clear manifestation of James Carey's "transportation as communication" idea. Now, observers were quick to make note of people throwing up, and pens that stopped working properly. But what are the other ramifications? Government officials were immediately lauding the rail line as a key to Tibet's development, while critics hailed it a threat to the Tibetan culture and environment. Both positions, no doubt, continue to ring true. We are seeing all kinds of new objects appearing in the world. And now much of the data exchange and processing duties being rerouted to the global communication satellite network accounts for most of the objects circling the world in geostationary orbit. This is Digination at the macro level. Of course, there is a micro-level story too.

In altering just slightly the wording in the following clause I've fashioned a call for a biological approach to understanding our Digination: "[w]hether or not [a biological approach] is true (either metaphysically or methodologically), or [just] a useful operating assumption, is an open empirical question and must be settled as such. The history of science tells us this is the only safe way to progress lest our theories become self-sealing" (Chemero and Silberstein, "Defending Extended Cognition," 2006, brackets added). Indeed, the "extended cognition" hypothesis is actually tightly bound up in a biological understanding of digital media use, so this isn't really much of a stretch at all.

A BIOLOGICAL APPROACH: SEEKING
THE SUSTAINABLE CYBORG

At the end of the first chapter, *Understanding our Digination*, I offered a brief sketch of the biological underpinnings of media ecology, the primary lens through which these different technological analyses were undertaken. An inductive approach guided most of these analyses. By employing relatively unobtrusive methods of observation I was able to glean fairly accurate and valuable insights into the meanings of my research participants' media selection, use, and behavior. While I admit making a few inferences along the way, the meanings of behaviors emerge primarily from participants' own actions in a web of interrelated activities and from their thoughts about those activities.

I think I successfully avoided committing any naturalistic fallacies that might inadvertently suggest that some more "natural" way of communicating (e.g., face-to-face conversation) is inherently good or necessarily better than some other means.

However, throughout the book, with this method in the background, I have been alluding to the need for a more *naturalistic* way of thinking about the human-technology relationship—about the human use of technology and about technology's use of humans. A biological metaphor (and perhaps more than just a metaphor) can go far in helping us see where we've been. It can also illuminate the current state of our affair with technology and even offer some perspective on where to go next.

My Mohawk friends do their best to take seriously the "seven-generation rule." From planting a crop, to buying a car, to sending a child off to college, to getting hooked up to the Internet, they try and look ahead to see what this decision might mean for the folks around at that time and place, and for people yet unborn, elsewhere. Like the Amish, though in a different way, they can be fickle decision-makers and pretty picky technology users. From the way we mine oil and coal to the way we install fiber-optic line, we may be reaching an impasse, or a tipping point, or even a point of no return with respect to our technology use today. This multi-generational way of thinking cannot remain on the fringes of mainstream culture any longer. And again, it is not just the technology users in question who this concerns.

Ecological economist Paul Ekins argues that "intergenerational equity" should be the guiding principle. According to Ekins, "present generations have an obligation to leave future generations no worse off in terms of environmental functions, in acknowledgement of the fact that these functions provide the basis for wealth-creation and economic activity, for human welfare and ways of life, and, indeed, for life itself" (in Constanza et al., 1996). We can and should interpret Ekins' concern for environmental functions to include our media environment as well. To again borrow from the psychologist J. J. Gibson, media are affordances in the environment that, in a sense, *want* to be used in certain ways. Different media represent different modes of awareness, different capacities for reflection, and therefore different functions of thought and forms of action. Despite the fact that the jettisoning of books has begun in earnest in libraries around the country, it is likely that the experience and outcome of reading Bronte or Dante, or Shakespeare or Woolf on a Kindle or iPad is not the same as that in a traditional codex. So in the same way we need to be vigilant with respect to preserving our bio-diversity, I want to argue for the preservation of our media diversity. Media are formative, they are consequential. Newer does not always mean better. Like the Mohawk, and maybe even the Amish too, we need to be picky media builders, growers, and users.

Without falling into any sort of naturalistic fallacy we can look closely at the natural world to see what really works well, and we can take good advantage of some natural tendencies in biological systems toward the production and recapitulation of organizational structures that seem to follow patterns of *balance* and *sustainability*. These patterns are the hallmarks of living things and so can offer some productive ways for us to relate to the tools we fashion and which, in turn, fashion us.

Traditionally, the sustainable relationship has been understood to mean that which is capable of continuing with minimal long-term negative effects on the environment. Concerning communication technologies, this might include informational, psychological, social, and political environments. It is not rare to hear someone say they would be incapacitated (or worse) if their cell phone, laptop, iPod, or Internet connection malfunctioned at the wrong time, or suddenly came up missing. If we can dismiss with most of the hyperbole in these testimonies, we can still locate some pretty unnerving examples of malfunction or loss. The point is that so many of us have grown very attached to and dependent upon our machines in a variety of psycho-physical ways (i.e., at intellectual, emotional, and/ or practical levels). To be sure, instances of communicative breakdown or loss can be severely debilitating in certain situations. But it is in looking closely at these instances that we begin to see the nature of the relationships maintained with and through the tools we use—and how the tools we use tend to find productive uses for us in return. I'll articulate what I mean by a sort of *purposive action* in tools and machines shortly to suggest how finding the right balance can be a tricky enterprise. First, however, we should consider some of the relevant history to get a better sense of where some significant imbalances in our human-technological systems have manifested themselves and taken hold.

About 1.5 million years ago the first group of proto-humans (probably homo-erectus) grasped the basics of catching, tending, and controlling fire. The controlled flame helped our ancestors begin to manage and direct this raw power of nature by first warming and illuminating an immediate area (extending the effect of skin and the acuity of the eyes). They also used fire to alter the chemical makeup of animal tissue and vegetable matter. Functionally speaking, this practice incorporates fire into the biological matrix. Combustion and cooking then forms an extension of the digestive system, so as to substantially broaden the diet. Today, the internal combustion engine is perhaps the most ubiquitous extension of the controlled flame, and it has become so much more. It is an extension of our own musculature and wider energetic systems. It's fair to say, we've gone a bit nuts with our cars and the petroleum fuels still lighting the fires that power them.

However, from an ecological or sustainability perspective this particular application of fire as an extension of human beings should have long ago been diminished or wholly discarded to make room for cleaner, quieter means of power and propulsion that showed promise. Unfortunately, without concerted and sustained effort on the part of public officials, industry leaders, and even the public to work toward seeing these alternative means developed, tried, and tested, they will always flounder and fail to "catch fire" due to a powerful kind of *network effect* that attends any broadly disseminated technological system or artifact. Successful adoption and diffusion of an innovation can simply be the by-products of the mass production, marketing, distribution, and cultural nurturing of just about any technological artifact, no matter how sensible or not-the-thing might be.

Indeed, the internal combustion engine today accounts for a lion's share of greenhouse gas and compound emissions worldwide. It also accounts for a lot of the sound pollution in urban areas, with the jet engine—a close relative to the internal combustion engine—producing the highest sound pressure levels most everywhere else. But we aren't the only ones who are noisy.

Andy Clark (2005) describes how the tiny mole cricket builds a cavity of a particular shape (similar to a Klipsch horn) in the ground that functions wonderfully as a sound amplifier and which enables the cricket, despite its diminutive dimensions, to become one of the loudest creatures on the planet. Clark argues that the cricket and the cavern become a third thing, a new organism, a bio-system with some impressive acoustic properties that give the little cyborg bug a leg up on the competition come mating season. In the process, of course, the now very extroverted mole cricket ensures the progression of this bio-technical strategy into the future. The cavern functions as part of the cricket's vocal apparatus. It is an extension of the being of the bug. An endless number of such interactions and integrations no doubt exist between organisms and their environments and between organisms too. The bee is the sexual organ of many species of flowering plants, the mother bird serves as a temporary part of the digestive system of her chicks. The bacterial colonies in the human gut are more permanent parts of our own digestive systems. Again, the list seems endless.

And so to get back to the human-machine scenarios, a stick becomes an extension of the arm. More intimate than fire, both organic and synthetic fibers function as extensions of our skin. Language sits in the middle of these designations—part nature, part culture—and it extends individual minds in powerful ways, but also persists across space and time to form a much broader, collective mind. And we should also consider in this way the many technological integrations that are more obvious or concrete: things like written language, photography, imagery, art, pocket notebooks, clocks, watches, alarms, calendars, barometers, thermometers, radios, web browsers, GPS systems, cell phones, iPods and all that. So many of these things have become part and parcel to us. They are often functioning as perfectly natural parts of our cognitive and mnemonic systems and in fact comprise new awareness and sensory arrays.

Given the way it has been applied, much of our current computer technology is in the process of moving to a level of hyper-utilization similar to that of the internal combustion engine. Take, for instance, the experience of the mobile technology user and all those who experience him. The idea is captured well enough in the concept of ubiquitous computing, whereby we enter a "computing environment in which each person is continually interacting with hundreds of nearby wireless interconnected computers" (Weiser, 1993; p.7). In 1993 when Mark Weiser first coined the phrase, ubiquitous computing was still very much the stuff of science fiction. Today, Weiser's vision has, more or less, come to fruition. In such an environment we find that many rules-of-thumb regarding technology use obsolesce, or are at least rendered nonsensical.

THE OBSOLESCENCE OF AUTONOMY AND SELF-REGULATION IN THE HUMAN-MACHINE COUPLING

Efficiency gurus are prone to suggest ways of managing our interactions with technologies of various kinds. New computer applications like *Freedom* and *Anti-Social* are designed to suspend access to things like Facebook, Twitter and even our favorite search engine. The aim of these new apps is to give us back our autonomy in order to help us think and get things accomplished. As funny as these developments may appear, and as talented as many of these designers are, such efforts miss something obvious about our relationships to these tools. It is that they are often very intimate and even intrinsic relationships. In other words, dedicated digital technology users can't really fathom the functional outcome of some sort of limited use. And so when the tech guru says "try to *go easy* on *this* or *that*," the advice really ends up being akin to saying "try to avoid using your eyes or ears." One piece of advice I've heard several renowned efficiency experts dole out at conferences or online concerns how we need to refashion our habits regarding e-mail. They suggest we get in the habit of checking e-mail just twice a day—perhaps three times at most. But with a smart phone on the hip or in one's pocket this ends up being a pretty silly piece of advice, for the device has become a natural and quite seamless extension of our mind-bodies. And this arrangement typically prompts perfunctory referencing, not at all unlike the way folks who still fancy watches absently glance at their wrists (and often without knowing what time it is when asked just seconds later). The automated suspension systems like *Isolator* and *Leechblock* work but don't seem much better as far as offering any sort of long-term fix. I predict they will be used by a very limited segment of the population because their widespread adoption seems antithetical to modern living if we look closely at the realities of the intimate relationships just described.

Consider the well-equipped teenager today who monitors their text messaging in a manner that would make the most acute obsessive compulsive feel in good company. A 2009 statistic from the *Pew Internet & American Life Project* reports that the average American teenager logs between 1500 and 3000 text messages a month. Unlimited text plans are certainly part of what spurs such numbers, and then habit takes over. When I undertook my own informal survey on the topic I discovered that the Pew numbers are probably conservative or possibly already out of date—or both. In the parking lot of my kid's school I get pretty nosey. And parents have disclosed to me the details of their Verizon and AT&T billing statements. I've discovered that 7000 to 9000 texts per month is not uncommon for many teenagers these days. With 6 to 8 hours of sleep built into the equation, this means a text is transmitted (sent or received) every 2 to 3 minutes, day in and day out. Despite what your favorite teenager will likely tell you, this is probably not a sustainable way of being, not psychically, socially, or emotionally. As I pointed out in an earlier chapter, we're finding that humans are notoriously bad multitaskers. Attention and listening skills are abysmal, and some very basic requirements that should probably attend a good and healthy existence are not being met. Some of these requirements certainly include adequate *presence* and participation in school and at work, in social situations, and in familial contexts. These parts of life have been altered by the thorough integration of texting (and devices supporting this practice) into the human attention-mnemonic-awareness system.

And with wireless hubs sprouting up in places as diverse as shopping malls, churches and synagogues, public beaches, bathrooms, taxicabs, and national forests, it's not just kids or folks with cell phones who are hyper-utilizing their technologies. It seems as though so many of us have little sense of when and where the signals should extend. My trusty devil's advocate asks: shouldn't they extend everywhere? Perhaps the question is already moot. In the context of Digination it has morphed into a kind of procedural dictate. This is why it's probably high time we take a closer look—*on the ground,* so to speak—at what's going on with our machines.

While I want to avoid pedaling the "viral" or "disease" analogy as I close this book because I don't think this is all quite as fatal as that, I am convinced that Digination is a sort of quasi-biological process run amok. And so we can ask again: should the extension of digital communication infrastructures and processes be so autonomous, so unlimited in their propagation and reach as to eventually blanket all the spaces and places of the world? This undoubtedly still sounds like a silly question to many of us already entwined, but I'm going to ask it just the same because it's the kind of question that rarely gets pondered by those in a position to make things happen (or not happen as the case may be). It all suggests that we are headed to and beyond the hyper-utilization of computer technology. I wondered in the first chapter whether Digination is like the weather. The answer I settled on was *sort of*, yet most technology users and policy-makers in positions to make a difference seem blissfully unaware of any distinction at all.

In remaining unaware we become incorporated and utilized by the same technologies we set in motion.[2] If not yet an instance of technology-out-of-control, it does appear, as Ralph Waldo Emerson once opined, that the things we use—all those objects and artifacts—are now "in the saddle riding us." In this way we become the conduits of commercial content, the manifestations, even the content itself, the sum and substance of corporate statistics and government databases. The system is out of balance and from a human perspective (which should mean a truly humane perspective) we find that this is not sustainable. An ecological outlook on our Digination prompts us to be much more conscious, imaginative, bold, and innovative in our efforts to maintain healthy, sustainable relationships with our machines. And despite continued assurances in certain quarters to the contrary (and even after the near-total collapse of the global economic system, our healthcare system, and our educational system[3]) it is clear that the logic of the marketplace won't help us find our way. The reason why the market can't guide us, with its supply-meeting-demand logic, is that it still *thinks* in monologic terms, with bottom lines, margins, and market shares dictating the course of technological innovation and diffusion. The market thinks, in other words, in terms of linear, one-way cause and effect.

As I was editing this final section of the manuscript in an attempt to mitigate the negative systemic effects of an editorial deadline already past, it was perfectly clear to me that while certain time pressures imposed (which, in fact, I imposed upon myself) were the consequence of digital technologies, these technologies are not the only problem. Some of the anxiety I was experiencing, and a general feeling of imbalance in my life at the time (and in the lives of my wife and children) also stemmed from certain exigencies associated with my career, from my own personal ambitions, as well as my own inabilities to manage time, personal stress, and familial responsibilities with grace and dignity.

So technology is not the only devil here. But electronic communication technology (and digital communication technologies in particular) can be catalysts to these imbalances, and the feeling of unsustainability so many of us detect. The "time sickness" I had succumbed to was often debilitating. I then tried editing this conclusion by hand during one session with pen and paper and it seemed more conducive to cognitive flow at the time because, among other things, it was also a lot easier on my back, wrists, and neck than some previous late-night editing sessions I had endured huddled over my laptop.

So if we certainly still need help dealing with time pressures and family pressures and work pressures and all that, we also need a new set of logics to help us navigate the future, to be more in the saddle of these new organisms we have become. But how should we characterize the relationships we currently maintain with our machines and systems? For starters, we need to characterize ourselves more in line with what we are—what we have always been. We become variants of something I call the *cyborg-symbiont-in-context.* We are hybridizations of human being and technological artifact. To be sure, we have always been this way. We don't just use technology, we intermingle and share and trade agency with our tools and machines.

Donna Haraway's (1991) image of a 1950s housewife pushing around a vacuum cleaner illustrates a legitimate point. There we see a good example of a cyborg. And she is more than a metaphor. There, at the surface, we see a machine that functionally extends the arms, hands, and fingers of the woman using the device. Of course the vacuum also enframes her as an efficient cleaning machine predisposed to doing such work. It sets up expectations and creates frustrations. And these together prompt some (many?) to seek out chemical palliatives to ease the stress, loneliness, and anxiety that result. Indeed, it prompts others to lash out at their children, and still others to leave the family context entirely. Haraway's image is a powerful one, but it might also prompt the reader to forget a great deal of our collective technological history and potentially confuses and occludes the special relationships humans have always had with their machines.

Indeed, without even straining the standard definition, we have been functioning as *cyborgs* for several million years. Our most ancient ancestors were attaching useful things to their bodies (first between thumb and forefinger), and later useful ideas and concepts to their minds (first with piles of stones or bones, or with fibers or sinews of various sorts, and then through early forms of art and language). These were some of our earliest technological appendages and extensions. In this book I considered the use of e-mail, blogs, iPods, laptop computers, smart phones, the Internet, search engines, and even *eBay* as mechanisms that append and extend us in all kinds of ways. One thing many of my research participants vividly revealed that so many of us are experiencing trouble avoiding the kind of perfunctory, hyper-utilization of our digital tools and machines described above. This has led to an imbalance in the way we sense and make sense of the world around us. We haven't always been so out of balance, but the increasingly ambiguous relationship between human and machine has stymied our efforts to locate sustainable examples of the cyborg-symbiont-in-context. I mentioned the sharing of intention and agency that has become the sine qua non of the digital age. In the next and final section I want to unpack some of what this sharing seems to entail.

DIGINATION AS A NATURAL BY-PRODUCT
OF MEDIA DESIGN LOGIC

The ecologist Edward Goldsmith (1997) notes that *purposiveness* is just another word for order or organization as applied to life processes. Goldsmith goes on to say that the two ideas are, in fact, inseparable. While the belief in various forms of consciousness, intelligence, and other forms of sentience in non-human entities is typical of advocates of the Gaia hypothesis, it is important to note that the notion of purposiveness does not and should not necessarily imply sentience, conscious intention, or intentional action. This reading of purpose is the feature most often cited and resisted by critics of broad systemic outlooks like Goldsmith's in the realm of natural ecology, or McLuhan's in the realm of media ecology.

One reason such critics tend to have so much trouble seeing the purposes or *roles* of individual entities is due to a kind of perceptual blindness that ends up being a side effect of any reductionist view of the world. Reductive perspectives, in their most pure and pernicious forms, tend to deride or dispose of any theory that conceives of events or processes in the world as "orderly and purposive rather than random, organized rather than atomized, co-operative rather than purely competitive, dynamic, creative, and intelligent rather than passive and robot-like, self-regulating rather than managed by some external agent . . . and tending to maintain stability or homeostasis rather than geared to perpetual change in an undefined direction" (Goldsmith, 1993; p. 5).

While the organization of most "primitive" societies depend on any number of animistic ontological frameworks, similar outlooks have lingered on the fringes of modern science as acceptable doctrine (and especially in ecology) for the past 150 years or so. It is not surprising to note, therefore, that in so much of the technology and media studies undertaken today, the idea of purpose-as-intention reduces to a kind of exogenous autonomous control that pits humans against the machines they build and use. However, and to reiterate once more, this general critique of something we might call *hard technological determinism* really is a kind of straw man argument marshaled in service of aging scientific outlooks bent on preserving their own worldview at the expense of more accurate representations of the world around us.

In much of the extant media studies, the demands, logics, biases, and other determining characteristics and features ascribed to various media continues to form a common misreading of media ecological arguments in general, and media or technological determinist arguments in particular. By assuming a primarily one-way relationship exists between media/technological systems and human organisms (with the human organism largely setting the terms of the relationship) we run the risk of missing the significance of very real pressures exerted back onto human users by the media and technologies that are deterministic cultural and social systems in and of themselves. Indeed, this is what Digination is all about.

Ostensibly, digital communication technologies enable a kind of near-instantaneous disembodied transportation, with fax machines, cell phones, e-mail and an Internet presence enabling people to get things accomplished near and far. These apparatuses can transport our voices, our thoughts, our signatures, and even our physical likeness when we cannot be (or do not wish to be) there in person. At one level, then, Digination is about the real-world transformations that are accompanying such virtual transportations. Of course, any effort to explain these processes is going to run into the problem of under-determination. We will never know *all* the connections. But we don't need to know everything.

The point is that thinking in ecological, even biological terms about the many communication technologies we regularly employ is a potent approach that has yet to be systematically applied to real-world media use. A media ecological perspective that takes its biological underpinnings seriously can help us understand processes of Digination to show how these tools function in altering the various ways people and societies come to *see* and *be* in their worlds. In short, a media ecological approach to understanding our Digination allows us to assess the fitness and predict the sustainability of digitally mediated relations, as well as the health and sustainability of the myriad social, educational, and political entities and institutions progressively finding themselves constituted by, if not actively constituting themselves through, such relations. Again, everything from anonymous professional collaborations, personal friendships, familial interactions, political parties, and national identities surely apply. We can find balance and communicative sustainability in each of these. An ecological approach can provide the guidance we need with respect to what should change and what should not in our efforts to improve ourselves as a species and, by extension, as a world society.

After literally hundreds of conversations with students, research participants, colleagues, and friends, I get the sense that most people, whether they've thought explicitly about it or not, seem to know intuitively that a biological approach to understanding media (and even the technology-as-organism idea) is a good idea, and one that isn't even much of a stretch. There are more than just surface similarities between things like the extensions and redirections of natural bodies of water with canals and artificial lakes, to our own nervous systems being enhanced with artificially flooded synapses teeming with serotonin, norepinephrine, and other neurotransmitters. From artificial nerves, organs, appendages and other bionic devices to the instantiation of a collective nervous system, Digination is a micro-to-macrocosmic story now happening all around and in us.

If we consider the practical relationships between people and things—the way each operates in the world, the nature/technology dichotomy loses much of its utility as an explanatory device. It becomes tenuous and even illusory. However, it appears as though the conveniences and tenacity of mass production and mass marketing have created a situation whereby our ability to know when enough communication processing speed and bandwidth really *is* enough has been, in a word, perturbed.

I think we have the means already in our midst for articulating and establishing an *eco-logic of media use* that can help us systematically assess the environmental and practical fitness of our new selves. We need to employ these means to make better sense of, and begin to guide the processes of, Digination. And this concerns much more than the way we use computers and other digital technologies. It also concerns the way we move about, the design of buildings, schools, cityscapes, and countrysides. Most of all, understanding our Digination concerns understanding the way we think.[4] Despite an emerging, if still mostly tacit cultural mandate that would suggest otherwise, we'll need expose our kids to many media—new and old.

NOTES

1. One good example to this point is *eBay* (described in "gory detail" in chapter 10). I'm not talking about artificial intelligence in the strict sense, or some occult notion of technological apparatuses exhibiting an autonomous sentience or volition of any kind. Nor am I referring to the effects of any values or biases of the designers of systems and networks built into those same systems and networks. Indeed, *eBay* is just one example of a global digital communication that is taking on a life of its own with the emergence of a committed community of regular—if not daily—users. The whole *eBay* structure requires the input and incorporates the content of monetary exchanges,

material culture, and the creativity, hopes, and deceits of human participants to make it *live*. *eBay* has become an economic, social, and political venue (and organism) all its own.

2. But this is to name just a few. We should also take a close ecological look at our disaster management and relief systems, our water system, our transportation systems, our energy production and distribution systems, our urban design and planning systems, etc.

3. While often organic or natural in their forms (and certainly reminiscent of biological structures), Geddes' and Corbusier's architectural and urban designs suggest at other times why natural shapes and systems are not always the best or most efficient in certain contexts. So natural logics or "bio-logics" can sometimes be trustworthy as guides in the move toward more human uses and scales. At other times they may not be as useful. For example, safe and efficient airplanes are not, ultimately as bird-like as da Vinci and other early inventors thought they could/should be. By the same token, and now concerning communication forms and modes, information theory (based on a kind of biological/nutrient transmission model) is not always the best model for human communication despite the focus on and push for efficiency since the telegraph came on the scene. In other words, perhaps we should not be following the dendritic tendencies of digital computers or the marketplace as mindlessly as we have thus far. Sometimes bio-mimesis doesn't work.

Appendix

The Tetrads

speed, contacts,
private, one-on-one
communication
(perception of)

carpal tunnel synd., eye trouble,
frustration, lost sense of time,
sleeplesness

self-awareness and
self-consciousness
in communication

solitude, hyper-self-consciousness
loss of identity (in the proxy)

control via hidden bureaucracy

memory

telegraphic style (texting...twitter...)

taking things literally

access to entire planet:
everybody, everywhere
(with an e-mail address)

info overload,
unwanted contacts, knowledge, etc.

message sent=message received (assumed)

E-mail

ENH | REV
RET | OBS

passing notes in class

precision,
editing, grammar, spelling,
punctuation, communicative
accountability

Tribal ecological sensibility:
knowing what's going on
trauma, paranoia*

conversation/F2F

telephone calls

text-based information
(the rational, the lineal):
Euclidean space,
Western time-space*

postage/stamps, postal system.

hand written correspondence,
hallmark cards,

History is what's in the
e-mail. Tribal memory
bank

See Tetrads: McLuhan and McLuhan (1988): radio*

The Hardcopy

283

The News Blog[1]
(technological level)

ENH | REV
RET | OBS

ENH (Enhances):

permalinks,
reverse-time
sequence:
"the latest is
the greatest"

hotlinks:
'diffusion' broadcasting:
the multilocational

perfect memory,
intellect:
always accessible,
in full

access to entire planet:
everybody,
everywhere*

REV (Reverses into):

Fundamentalism
(as the "corporate"
counter-reacts   against the chaotic
surplus of info)

Anarchism
("info wants to be free")

"plants," subterfuge, hidden bureaucracy**

text → imagery: (blogs to vlogs)

Global Village Story-Time~*:
story-tellers participating in their own audience participation

precision,
editing, grammar, spelling,
accountability,
formal vetting process

Network News

Top-down, corporate, centralized news
hierarchy (and its apparatus') may implode,
invert, and ultimately scatter.

RET (Retrieves):

tribal ecological sensibility:
trauma, paranoia*

pamphleteers:
not pamphlets themselves,
but the people and the act
of pamphleteering

OBS (Obsolesces):

text-based information
(the rational, the lineal):
Euclidean space,
Western time-space*

"Orality" (secondary)

History is what We make it.
Blog as tribal memory
bank***

See tetrads:* radio, **computer,

the undefined crowd ; conformity and,
multiple points of view at once, vertigo, confusion
(blogger in Madrid pushing union activism in Detroit)

"Global (Village) Story Time"***
World reverses into script: contributors are story-tellers
participating in their own audience participation

Self-consciousness,
individualism (feeling of), the ego

national public spheres

commitment

private, inner life

perspective,
point-of-view*

The Present
and (hopeful) Future:
the "Dream of Modernity"
(newer means better)via
time-stamp

detachment,
objectivity*

$$\begin{array}{c|c} \text{ENH} & \text{REV} \\ \hline \text{RET} & \text{OBS} \end{array}$$

mosaic image
of the public*

modernity, rationality

The News Blog²
(cultural/socio-political level)

the intellectual (idea of)
as an important figure (the blog editor:
interpreter and developer of information
filters for the population),

language as aesthetic
object

communal
awareness as
point of
view

See Tetrads: *Telegraph,
**Radio, ***visual space

The Internet

ENH | REV
RET | OBS

face in the crowd

solitude, intolerance,
impatience, neuroses

hassles, frustration, identity theft,
violence from strangers

Imagery, Right Hemispheric Bias

control via hidden bureaucracy
embedded in the interface

carpal tunnel, eye trouble, back trouble, obesity

lost sense of both time and physical space

sleeplessness

infophilia, information overload,
unwanted information, experience and awareness

Point of view,
sense of place,

conversation/F2F,

telephone calls,

Network News:
The Anchorman
"The way it is"
Authority

The Word:
books, libraries, bookstores,
schools and universities

Reading/Literacy
Left Hemisphere

all info all the time,
omniscience, ESP

Intellect, Knowledge, info, data
(collapse and collusion of)

Education (access to)

Connection, visibility

Self-expression

Transactions… Business

access to entire planet:
everything, everywhere
(on the Internet)

Tribal ecological sensibility:
emotional response,
knowing what's going on, gossip,
trauma, paranoia*

an eye for an ear

the Village

familiar face

Proximity to others

Right Hemisphere

See tetrads:* radio, **computer,

Google

ENH | REV
RET | OBS

omniscience, ESP

Intellect, Knowledge,

Connection, visibility

Transactions... Business

access to entire planet:
everything, everywhere
(on the Internet)

Tribal ecological sensibility.
emotional response,
knowing what's going on, gossip,
trauma, paranoia*

an eye for an ear

the enclosure

the Citadel

The "map"
(is not the territory)

Popularity Contest

propaganda, engine gaming,
"plants"

control via hidden bureaucracy
embedded in the algorithm

infophilia, information overload,
unwanted information, experience and awareness

Point of view,
sense of place,

conversation/F2F,

The Research Enterprise

Network News:
The Anchorman
"The way it is"
Authority

The Word:
books, libraries, bookstores,
schools and universities

Reading/Literacy
Left Hemisphere

See tetrads:* radio, **computer,
***credit card, and Internet** ...

The Personal Digital Music Device

ENH | REV
RET | OBS

ENH

" planned spontaneity: " random/shuffle feature and massive memory capacities in PDMDs allow for controlled de-control of the emotions.

aesthetic experience "All art [life], aspires toward the condition of music"

autonomy, libertarianism, and ego

privacy

REV

Autonomy reverses into Automatony. "I'm an emotional mannequin")

Privacy reverses into civic "free rider" scenario. Individual becomes "social minimizer."

world reverses into virtual reality: "toy domain"

social displacement: "all alone together" but nowhere in particular. People disappear as interacting subjects. "I am not 'there'."

solipsism

RET

part of tribal ecological sense (spontaneous emotional response): elation, anger, calm, trauma, paranoia "it really makes me feel alive"

anthropomorphism: "my iPod proved that it is the smartest person I know, and that it has a sick sense of humor and impeccable taste in women."

OBS

hearing, listening to, talking with strangers

social interaction

the "Agora"

the citizen

democracy

civic sphere dissolves out of public space

polity collapses

... the album

The Mobile Podcast

Tetrad center:

ENH | REV
RET | OBS

Enhances (Percepts (and primacy of appearances)):

Self-Centeredness and Predisposition (over sound, space, emotion, time, etc.). " planned spontaneity:" (Ex: Ingraham and Hayworth) controlled de-control of the emotions** innuendo, opinion

Speech (as aesthetic experience) "Participation" in the "News"

"Glocalization" and virtual social placement: Disconnected Tribes "all alone together"

Publicy** and the Parasocial Relation

Retrieves (Tribal Sensibility*):

(collectivism, outer-directedness, and spontaneous emotional response): suspicion, elation, anger, calm, trauma, paranoia, etc.

the circumstantial, happenstance: "I never miss the sermons" "I notice all these new things…" "here and now"

Language as aesthetic object*** (via cliché, metaphor, allusion, aphorism, turns of phrase)

hearsay, gossip Soothsayer, Seer, Wise One

Communication as Ritual

Reverses into (Virtual Tribal Space.):

Autonomy reverses into Automatony (consumer's experience often explicitly prompted even interpreted by decontextualized narrator/s).**

Privacy (arguments, phone conversations, wonderings, etc) reverses into Publicy

physical /social environment reverses into story prop: "all the world's a stage"

Anarchy (Wise One reverses into Lunatic)

Primary Orality becomes Secondary Orality

Collectivism reverses into Solipsism **

Groupthink (turns to vitriol and polarization

Obsolesces (Concepts: abstract, linear, logical, sequential thoughts and utterances):

Reason, Rationality, "objective" news

social interaction, discourse, talking with strangers** the local "Agora"**

geographic communities, neighborhoods

the embodied citizen

Deliberative Democracy** civic sphere dissolves out of public space, **

freedom of thought pluralsim

polity

Tetrads: McLuhan and McLuhan (1988): radio*; MacDougall (2005, 2005); Personal Digital Music Device**, Blogs***

The W²M²D²*

(ENH | REV / RET | OBS)

(*WIFI/Wireless Multi-Mediating Digital Device)
(sometimes AKA Weapons of Mass Distraction)

face in the crowd
solitude, intolerance,
impatience, neuroses
all alone together, ADD,
ants-in-the-pants, mental
wandering, knee-jerk response,
panopticon
hassles, frustration, identity theft,
violence from strangers

imagery, Right Hemisphere

control via hidden bureaucracy embedded in interface

carpal tunnel synd., eye trouble, back trouble, obesity

lost sense/awareness of time and physical space, sleeplessness

infomania, infophilia, infotropism, information overload,
unwanted information, experience and awareness

Point of view,
sense of place,

privacy, security

conversation/F2F,
telephone calls,

Network News:
The Anchorman
"The way it is"
Authority
sense of place, classroom, questions,
contemplation, mental wondering,
class participation, the timid
The Word: books, libraries,
bookstores, schools and
universities

Reading/Literacy
Left Hemisphere

All info all the time,
omniscience, ESP

astral projection, ego,
self-centeredness,
room with a view, class participation,
boldness

Intellect, Knowledge,

Education (access to)

Connection, visibility,
Self-expression

Transactions... Business

access to entire planet:
everything, everywhere
(on the Internet)

Tribal ecological sensibility:
emotional response,

knowing what's going on, gossip,
trauma, paranoia

an eye for an ear

the Village

familiar face

"Proximity" to others

little secrets, gossip,
class participation

References

Aday, S (2005). "The Real War Will Never Get on Television: An Analysis of Casualty Imagery in American Television Coverage of the Iraq War," in *Media and Conflict in the Twenty-First Century*, ed. Philip Seib, New York, NY: Palgrave MacMillan, pp. 141–155.

Agosto, N (2002). "Bounded Rationality and Satisficing in Young People's Web-based Decision Making." In *Journal of the American Society for Information Science and Technology*. Volume 53, I; pp. 16–27

Alexander, Joy (2005). Reading with the Ear: the Necessity for Language and Literacy Learning of the Aural Imagination. www3.educ.sfu.ca/conferences/ierg2005/viewpaper.php?id=61andprint=1.

Altheide, D. L., and Snow, R. P. (1979). *Media Logic*. Beverly Hills, CA: Sage.

Ashbrook, T. (2009). *On Point*, segment 1. "Multitasking Minds." September 29.

Aspden, P (2005). Me, Myself, iPod. news.ft.com/cms/s/4bc9ffdc-4c13–11da-997b-0000779e2340.html.

Attali, J. (1985) *Noise: The Political Economy of Music*. Minneapolis: University of Minnesota Press.

Avery, R. K., and McCain, T.A. (1982). "Interpersonal and Mediated Encounters: A Reorientation to the Mass Communication Process." In G. Gumpert and R. Cathcart (eds.), *Inter/Media: Interpersonal Communication in a Media World*. New York, NY: Oxford University Press.

Ballard, J. G. (1997). *A User's Guided to the Millenium*. Fort Lauderdale, FL: Flamingo Press.

Baudrillard, J. (1988*)*. "Simulacra and Simulations." In *Selected Writings,* ed. Mark Poster. Palo Alto, CA: Stanford University Press, pp.166–84.

Bavelas, J. B., Hutchinson, S., Kenwood, C., and Matheson, D. H. (1997). "Using Face-to-Face Dialogue as a Standard for Other Communication Systems." *Canadian Journal of Communication, 22*, pp. 5–24.

Bench-Capon, T. J., and McEnery, A. M. (1989). "People Interact through Computers Not with Them." *Interacting with Computers, 1*, 1.

Benedetti, P., and DeHart, N. (1996). *Forward through the Rearview Mirror: Reflections on and by Marshall McLuhan.* Cambridge, MA: MIT Press.

Benkler, Y (2006). *The Wealth of Networks: How Social Production Transforms Markets and Freedom.* New Haven, CT: Yale University Press.

———— (2006a). Speaking on the Wealth of Networks. Tuesday, April 18, at 5:45 pm in 102 Hauser Hall at Harvard Law School. Audio available at: media-cyber.law .harvard.edu/AudioBerkman/old/12/58/benkler_2006–04–24.mp3.

Bennett, W. L. (2003). "Communicating Global Activism: Strengths and Vulnerabilities of Networked Politics." *Information Communication and Society, 6*(2), pp. 143–168.

———— (2005). "News as Reality TV: Election Coverage and the Democratization of Truth. *Critical Studies in Mass Communication, 22*(2), pp. 171–77.

Berger, P., Berger B., and Kellner, H. (1973). *The Homeless Mind: Modernization and Consciousness.* New York, NY: Vintage.

Berners-Lee, T. (2006). *Neutrality of the Net.* Decentralized Information Group (DIG). Submitted May 2 dig.csail.mit.edu/breadcrumbs/node/132 (Retrieved December 18, 2006).

———— (2007). "Welcome to the Semantic Web." In *The Economist: World in 2007*, print edition. Available online at www.economist.com/theworldin/science/ displayStory.cfm?story_id=8134382In

Birdwhistell. R. L. (1970). *Kinesics and Context.* Philadelphia, PA: University of Pennsylvania Press.

Blogging and Publicy. (2003, December 19). *What is the Message?* Retrieved from the McLuhan Program. Web site: www.mcluhan.utoronto.ca/blogger/2003_12_01_ blogarchive.html#107184093362428431.

Boczkowski, P. J. (1999). "Mutual Shaping of Users and Technologies in a National Virtual Community." *Journal of Communication. 49.* pp. 86–108.

Boettinger, H. M. (1977). *The Telephone Book: Bell, Watson, Vail and American Life, 1876–1976.* Croton-on-Hudson, NY: Riverwood Publishers.

Bolter, J. David, and Grusin, Richard (1999). *Remediation: Understanding New Media.* Cambridge, MA: MIT Press.

Boorstin, D. (1961). *The Image: A Guide to Pseudo-Events in America.* New York, NY: Vintage.

Bower, B (1999). "Simple Minds, Smart Choices, for Sweet Decisions, Mix a Dash of Knowledge with a Cup of Ignorance." www.sciencenews.org/pages/sn_arc99/ 5_29_99/bob2.htm.

Bowers, C. A. (2000). *Let Them Eat Data: How Computers Affect Education, Cultural Diversity, And Ecological Sustainability.* Athens, GA: University of Georgia.

Buber, Martin (1977). *I and Thou.* Beaverton, OR: Touchstone Free Press

Budzik, J., and Hammond, K. (1999). "Watson: Anticipating and Contextualizing Information Needs." In *Proceedings of the Sixty-second Annual Meeting of the American Society for Information Science.* citeseer.ist.psu.edu/budzik99watson.html.

Bull, M. (2000) *Sounding Out the City: Personal Stereos and the Management of Everyday Life*. Oxford, UK: Berg.

Burke, K. (1966). Language as Symbolic Action: Essays on Life, Literature, and Method. Berkeley, CA: University of California Press.

Byrne, B. (2005). The iPod: Development or Digestion? *aliasfrequencies.org/bb/the-ipod-development-or-digestion.*

Carbaugh, D. (1988). *Talking American: Cultural Discourses on Donahue*. Norwood, NJ: Ablex.

_____ (1990). *Cultural Communication and Intercultural Contact*. Albany, NY: SUNY Press.

_____ (1996). *Situating Selves: The Communication of Social Identities in American Scenes*. Albany, NY: SUNY Press.

Carey, J. (1989). *Communication as Culture*. New York, NY: Routledge.

Carr, Nicholas (2008). *Is Google Making us Stupid?* www.theatlantic.com/doc/200807/google.

Carpini, M. (1996).Voters, Candidates, and Campaigns in the New Information Age: An Overview and Assessment. *Harvard International Journal of Press/Politics*, *1*, pp. 36–56.

Chaffee, S. (1982). "Mass Media and Interpersonal Channels." In Gumpert and Cathcart, *Inter/Media: Interpersonal Communication in a Media World*. New York, NY: Oxford University Press.

Chesebro, J., and Bertelson, D. (1996). *Analyzing Media: Communication Technologies as Symbolic and Cognitive Systems*. New York, NY: The Guilford Press.

Chomsky, N. (2001). *9–11*. New York, NY: Seven Stories Press.

Clark, A. (1997). *Being There: Putting Brain, Body and World Together Again*. Cambridge, MA: MIT Press.

_____ (1998)

_____ (2003). *Natural-Born Cyborgs: Minds, Technologies and the Future of Human Intelligence*. New York, NY: Oxford University Press.

Chalmers, C., and Chalmers, D. (1998) *The Extended Mind*. St. Louis, MO: Washington University.

Chronicle (2009). Professor Encourages Students to Pass Notes During Class—via Twitter chronicle.com/wiredcampus/article/3705/professor-encourages-students-to-pass-notes-during-class-via-twitter.

Constanza, R., Segura, O., and Martinez-Alier, J. (1996). *Getting Down to Earth: Practical Applications of Ecological Economics*. Washington, DC: Island Press.

Csikszentmihalyi, M. (1990). *Flow: the Psychology of Optimal Experience*. New York, NY: Harper Perennial.

Culnan, M. J., and Markus, M. L. (1987). Information Technologies. In F.M. Jablin, L.L. Putnam, K.H. Roberts, and L.W. Porter (eds.), *Handbook of Organizational Communication: An Interdisciplinary Perspective* (pp. 420–443). Newbury Park, CA: Sage.

Debord, G. (1981). *"Theory of the Dérive."* In *Situationist International Anthology*. Berkeley, CA: Bureau of Public Secrets.

De Certeau, M. (1984). *The Practice of Everyday Life*. Steven Rendall, trans. Berkeley, CA: University of California Press.

Deenan, S. (1991). "Doing the Boca: an Interim Report from a Reinvented Newspaper." *Columbia Journalism Review*. May/June.

DeSanctis, G. and Gallupe, R. B. (1987). A Foundation for the Study of Group Decision Support Systems. *Management Science, 33*, pp. 589–609.

Desmond, J. (2003). *Consuming Behavior*. London, UK: Palgrave.

Deuze, M. (2003). "The Web and its Journalisms: Considering the Consequences of Different Types of Newsmedia Online." *New Media and Society, 5*(2), pp. 203–30.

Dolby, N. (2000). "The Tyranny of the Overhead: the Conference Paper Reconsidered." Teachers College Record Date Published: 4/5/00. Located at www.tcredord.org (ID Number: 10515, Date Accessed: 6/21/00).

Donaldson, G. (1997). *Abundance and Anxiety: America, 1945–1960*. New York, NY: Praeger.

Du Gay, P. and Hall, S. (eds.) (1997) *Doing Cultural Studies. The Story of the Sony Walkman*. London, UK: Sage.

Dyson, G. (1997). *Darwin Among the Machines: the Evolution of Global Intelligence*. New York, NY: Perseus Books.

The Economist, 375. no. 8430 (June 11, 2005): p. 9.

"*eBay Facts, Growth Statistics and FAQs*" (2005) www.thebidfloor.com/ebay_facts.htm (retrieved April 18, 2008).

Eisenstein, E. (1979). *The Printing Press as an Agent of Change: Communications and Cultural Transformations in Early-Modern Europe*. New York, NY: Cambridge University Press.

Ellul, J. (1964). *The Technological Society*. Trans. John Wilkinson. New York, NY: Knopf.

Eriksen, T. H. (1996). *The Internet, the "Laws of Media" and Identity Politics*. Retrieved June 16, 2005, from folk.uio.no/geirthe/Tetrads.html

Fallows, J. (2005). "A Journey to the Center of Yahoo." *New York Times Online*. Techno Files. November 6. www.nytimes.com/2005/11/06/business/yourmoney/06techno.html).

Federman (2003). *Publicy*. www.mcluhan.utoronto.ca/blogger/2003_12_01_blogarchive.html#10718409336.

Ferguson, C. H. (2005). "What's Next for Google: Running the Web's Best Search Engine isn't Enough. In *Technology Review*. January. www.technologyreview.com/ InfoTech/wtr_14065,308,p1.html.

Feenberg, A. (1993). Building a Global Network: The WBSI Experience. In L. Harasim (ed.), *Global Networks* (pp. 185–197). Cambridge, MA: MIT Press.

Fish, S. (1980). *Is There a Text in this Class?* Cambridge, MA: Harvard University Press.

Foucault, M. (1973). *Birth of the Clinic*. New York, NY: Pantheon
_____ (1976). *Discipline and Punish*. New York, NY: Pantheon.

Fox, B. (2005). "Google searches for quality not quantity." Special Report from *New Scientist*. April 30.www.newscientist.com/channel/info-tech/mg18624975.900

French, G. (2004). "Yahoo! vs. Google: Algorithm Standoff." *Web Pro News.* www.webpronews.com/insiderreports/searchinsider/wpn-49-20040224 YahooVsGoogleAlgorithmStandoff.html

Gabardi, W. (2000). *Negotiating Postmodernism.* Minneapolis, MN: University of Minnesota Press.

Gant, M. (2006). Someone's Listening. www.metroactive.com/metro/06.07.06/listening-post-0623.html

Gardner, H., and Hatch, T. (1989). Multiple Intelligences go to School: Educational Implications of the Theory of Multiple Intelligences. *Educational Researcher, 18*(8), 4–9.

Gibson, J. J. (1966). *The Senses Considered as Perceptual Systems.* Boston, MA: Houghton Mifflin.

Gibson, J. J. (1979). *The Ecological Approach to Visual Perception.* Boston, MA: Houghton Mifflin

Gibson, W. (1993). "Disneyland with the Death Penalty. In *Wired.* Issue 1.04 Sep/Oct.

Gitlin, T. (1980). *The Whole World is Watching: Mass Media in the Making and Unmaking of the New Left.* Berkeley, CA: University of California Press.

Goffman, E. (1967). *Interaction Ritual: Essays on Face-to-Face Behavior.* Garden City, NY : Doubleday and Co.

_____ (1973). *The Presentation of Self in Everyday Life.* Woodstock, NY: Overlook Press.

Goldsmith (1997). "*The Way: an Ecological World View.*" *InterCulture* Vol. 30 no. 1, Winter–Spring.

Granovetter, M. S. (1973). The Strength of Wweak Ties. *American Journal of Sociology, 78,* pp. 1360–1380.

Gross, T. (2008) with special guest Joseph Stiglitz: "Our 'Three Trillion Dollar War.'" *Fresh Air.* WHYY. March 3.

Grossman, L., and Hamilton, A. (2004, June 13). Meet Joe Blog. *Time.* Retrieved from www.time.com/time/magazine/article/0,9171,1101040621–650732–1,00.html

Greenfield, P. (1993). Representational Competence in Shared Symbol Systems: Electronic Media from Radio to Video Games. In R.R. Cocking, and LA. Renninger, *(8th.) The Development and Meaning of Psychological Distance* (pp. 161–183). Hlllsdale, NJ: Erlbaum.

Gumpert, G., and Cathcart, R. (1982). *Inter/Media: Interpersonal Communication in a Media World.* New York, NY: Oxford University Press.

_____ (1985). Media Grammars, Generations, and Media Gaps. *Critical Studies in Mass Communication,* 2, pp. 23–35.

Hatch, T and Gardner, H. (1993) 'Finding Cognition in the Classroom . . .' in G. Salomon (ed.) *Distributed Cognitions. Psychological and Educational Considerations,* Cambridge, MA: Cambridge University Press.

Hatchen, W. and Hatchen, H. (1992). "Reporting the Gulf War," in *The World News Prism: Changing Media of International Communication,* 3rd. ed. Ames, IA: Iowa State University press.

Haraway (1991). *Simians, Cyborgs, and Women: The Reinvention of Nature.* New York, NY: Routledge, and London: Free Association Books.

Heidegger, M. (1977). *The Question Concerning Technology and Other Essays.* New York, NY: Garland Publishing.

Hegel, G. W. F. (1977). *The Phenomenology of Spirit* (A. V. Miller, Trans.). New York, NY: Oxford University Press.

Herman, E. S., and Chomsky, N. (1988). *Manufacturing Consent: the Political Economy of the Mass Media.* New York, NY: Pantheon Books.

Hillis, Petit, and Epley (2006). *Everyday eBay: Culture, Collecting, and Desire.* New York, NY: Routledge.

Hobbes, Thomas (1950). *Leviathan.* New York, NY: Dutton.

Hodgson, G. (2000). The End of the Grand narrative and the Death of News. *Historical Journal of Film, Radio and Television,* March; v20, p. 23.

Horton, D. and Wohl, R. (1954). Mass Communication and Para-Social Interaction: Observations on Intimacy at a Distance. *Psychiatry, 19,* pp. 215–29.

Hosokawa, S. (1984) *'The Walkman Effect,'* Popular Music 4: pp. 165–80.

Hourihan, M. (2002). *What We're Doing when We Blog.* Retrieved from www.davenet.scriptng.com/2002/07/31/megHourihanWhatWereDoingWhenWeBlog

Howard, P. J. (2006). *The Owner's Manual for the Brain.* 3rd edition. Austin, TX: Bard Press.

Husserl, E. (1936/1970). *The Crisis of European Sciences and Transcendental Phenomenology.* Chicago, IL: Northwestern University Press.

Innis, H. (1950). *Empire and Communications.* Oxford: University of Oxford Press.

———— (1951). *The Bias of Communication.* Toronto, ON: University of Toronto Press.

International Trade Statistics Yearbook (2000), Volume I. Trade by Country, Volume II. Trade by Commodity, Series: G, No.49

iPods as a "cocoon of solipsism." (2005). *Discourse.net.* Retrieved from www.discourse.net/archives/2005/06/ ipods_as_a_cocoon_of_solipsism.html

James, W. (1890/1950). *The Principles of Psychology,* Vol. 1. Mineola, NY: Dover Publications.

Johnson, S. (1997). *Interface Culture: How New Technology Transforms the Way We Create and Communicate.* San Francisco, CA: Harper Edge.

Katz, E. and Lazarsfeld, P. (1955). *Personal Influence.* New York, NY: The Free Press.

Katz and Katz (1998). "McLuhan: Where Did He Come From, Where Did He Disappear?" *Canadian Journal of Communication.* Vol. 23, No 3.

Katz, L. (1953). "A New Status Index Derived from Sociometric Analysis." *Psychometrika,* 18: pp. 39–43.

Keepnews, J. (1997, August 14–20). Screen Angst. *Metroland,* p. 11.

Kiesler, S., Siegel, J., and McGuire, Y. W. (1984). Social Psychological Aspects of Computer-Mediated Communication. *American Psychologist,* 39(10), pp. 1123–1134.

Kilborn, P. (1990). Workers Using Computers Find a Supervisor Inside. *New York Times,* December 23, p. 1.

Kim, T. and Biocca, F. (1997). Telepresence via Television: Two Dimensions of Telepresence may Have Different Connections to Memory and Persuasion. *Journal of Computer-Mediated Communication*, *3*(2). Retrieved from jcmc.indiana.edu/vol3/issue2/kim.html.

Kleinberg, J. (1999). "Authoritative Sources in a Hyperlinked Environment." *Journal of ACM* (JASM), 46, 1999.

Kleinberg, J. and Raghavan, P. (2005). "Query Incentive Networks." In proceedings of *IEEE Symposium on Foundations of Computer Science*, Pittsburgh, PA, October 23–25.

Kracauer, S. 1995. *The Mass Ornament: Weimar Essays*. 3rd Edition. London, UK: Harvard University Press.

LaLand, K. N. and Brown, G. R. (2006). Niche Construction, Human Behavior, and the Adaptive Lag Hypothesis. *Evolutionary Anthropology, 15,* pp. 95–104.

Latour, B. (1995). "Mixing Humans and Nonhumans Together: The Sociology of the Door-Closer." In: S.L. Star (Ed.), *Ecologies of Knowledge: Work and Politics in Science and Technology*. Albany, NY: SUNY Press.

Lea, M., and Spears, R. (1995). Love and First Byte? Building Personal Relationships over Computer Networks. In J. T. Wood and S. Duck (eds.), *Understudies Relationships: Off the Beaten Track.* Thousand Oaks, CA: Sage.

Linden, G. (2004). "Humans vs. Robots = Google vs. Yahoo." glinden.blogspot.com/2004/09/ humans-vs-robots-yahoo-vs-google.html.

Logan, R. K. (2007). *The Extended Mind: The Emergence of Language, the Human Mind and Culture.* Toronto, ON: University of Toronto Press.

Lum, C. M. K. (1996). *In Search of a Voice: Karaoke and the Construction of Identity in Chinese America.* Hillsdale, NJ: Lawrence Erlbaum.

Lyons, K. (2004). As Laptop Use Rises, so Does Virtual Note Taking. www.softwaresecure.com/pdf/Dartmouth_Article.pdf

MacDougall, R. C. (1999) "Subject Fields, Oral Emulation, and the Spontaneous Cultural Positioning of Mohawk e-mail Users" in *World Communication, 4,* 28; pp. 5–25.

_____ (2001). *Electronic Mail at a Mohawk Indian Community: an Investigation into Meaning and Use.* Ann Arbor, MI: University of Michigan Press.

_____ (2005). "Identity, Electronic Ethos and Blogs: a Technologic Analysis of Symbolic Exchange on the New News Medium." *American Behavioral Scientist.* Vol. 49, No. 3.

_____ (2008). "Web Search Engineering and our Emerging Information Ecologies: Designing and Defining Authority Online." *Explorations in Media Ecology.* Vol. 7, No. 2.

_____ (2009). Cultural Technics: Making Meaning at the Interfaces of Oral and Electronic Culture. Verlag VDM.

_____ (2010). "eBay Ethics: Simulating Civility Today for the 'Digital Democracy' of Tomorrow." *Convergence: The International Journal of Research into New Media Technologies.* May. Vol. 16 No. 2. pp. 235–244.

McCarthy, J. and Hayes, P. J. (1969). "Some Philosophical Problems from the Standpoint of Artificial Intelligence," in *Machine Intelligence 4*, ed. D. Michie and B. Meltzer, Edinburgh, UK: Edinburgh University Press, pp. 463–502

McClamrock, R. (1995). *Existential Cognition: Computational Minds in the World.* Chicago, IL: Chicago University Press.

McKenna, K. (1999). "Missed Story Syndrome," in *Impact of Mass Media*, ed. Ray Eldon Hiebert. New York, NY: Longman.

McLuhan, M (1951). *The Mechanical Bride: Folklore of Industrial Man*. New York, NY: Vanguard Press.

_____ (1955). *New Media as Political Forms*. Originally published in Explorations 3. To be republished in: *Marshall McLuhan Unbound 1:* Gingkopress (www .gingkopress.com/_cata/_mclu/_newmed.htm).

_____ (1962). *The Gutenberg Galaxy: The Making of Typographic Man*. Toronto, ON: University of Toronto Press.

_____ (1964). *Understanding Media: The Extensions of Man*. New York, NY: McGraw-Hill.

_____ (1977). Laws of Media. *English Journal*, *67*(8), 92–94.

_____ (1997). *Forward Through The Rearview Mirror*. Cambridge, MA: MIT Press.

McLuhan, M., Fiore, Q., and Agel, J. (1967). *The Medium is the Massage: An Inventory of Effects*. New York, NY: Bantam Books.

McLuhan, M. and McLuhan, E. (1988). *Laws of Media: The New Science*. Toronto, ON: University of Toronto Press.

McLuhan, M. and Powers, B. B. (1989). *The Global Village: Transformations in World Life and Media in the 21st Century*. New York, NY: Oxford University Press.

McWilliams, G. (2005). "Laptops in Classrooms not Working Out as Hoped." *Wall Street Journal*. Oct 14.

Mead, G. H. (1934). *Mind, Self and Society*. Chicago, IL: University of Chicago Press.

Merleau-Ponty, M. (1945, 2005). *Phenomenology of Perception*. Trans: Colin Smith, London, UK: Routledge.

Metro (2007). Boston edition. June 9–12.

Meyrowitz, J. (1985). *No Sense of Place: The Impact of Electronic Media on Social Behavior*. Oxford, UK: Oxford University press.

_____. (2005). *The Rise of Glocality: New Senses of Place and Identity in the Global Village* 21st.century.phil-inst.hu/Passagen_engl4_Meyrowitz.pdf

Mitroff, I. A., and Bennis, W. (1989). *The Unreality Industry: the Deliberate Manufacturing of Falsehood and What it Is Doing to Your Lives*. New York, NY: Oxford University Press.

Moebius, H. and Michel-Annen, B. (1994) *Colouring the Grey Everyday: The Psychology of the Walkman*. Free Associations 4/4(32): pp. 570–77.

Montaigne, M. (1550). "Of the Inconstancy of our Actions." In *The Essays of Michel de Montaigne* (2005). Whitefish, MT: Kessinger Publishing.

Mumford, L. (1934). *Technics and Civilization*. New York, NY: Harcourt, Brace and Co.

Myers, D. (1987). "Anonymity is Part of the Magic": Individual Manipulation of Computer-Mediated Communication Contexts. *Qualitative Sociology*, *10*, pp. 251–266.

Myles, J. (2004). Community Networks and Cultural Intermediaries: The Politics of Community Net Development in Greater Manchester. *Media, Culture and Society*, 26(4), pp. 467–90.

Negroponte, N. (1995). *Being Digital*. New York, NY: Knopf.

Nideffer, R. (1996). A Book Review of Rosanne A. Stone's *The War of Desire and Technology at the Close of the Mechanical Age*. Social Science Computer Review, Vol. 14, No. 4, pp. 483–87

Nietzsche, F. (1990). *The Genealogy of Morals*. New York, NY: Anchor.

Nilsen, G. (2003). "Analysis of Link Structures on the World Wide Web and Classified Improvements." Pittsburgh: University of Pittsburgh. April.

Norris, P. (2004). The Bridging and Bonding Role of Online Communities. In P. Howard and S. Jones (Eds.), *Society Online: The Internet in Context* (pp. 31–44). Thousand Oaks, CA: Sage.

Nystrom, C. (1973). *Towards a Science of Media Ecology: The Formulation of Integrated Conceptual Paradigms for the Study of Human Communication Systems*. Doctoral Dissertation, New York University.

Odling-Smee F .J., Laland, K. N., and Feldman, M. W. (2003). *Niche Construction: The Neglected Process in Evolution*. Princeton, NJ: Princeton University Press.

Oldenburg, R. (1999). *The Great Good Place: Cafes, Coffee Shops, Bookstores, Bars, Hair Salons, and Other Hangouts at the Heart of a Community*. New York, NY: Marlowe.

Ong, W. J. (1980). Literacy and Orality in Our Times. *Journal of Communication*, 30, pp. 197–204.

_____ (1982). *Orality and Literacy: The Technologizing of the Word*. London, UK: Routledge.

Ophir, E., Nass, C., and Wagner, E. (2009). *Cognitive Control in Media Multitaskers. Proceedings of the National Academy of Sciences*. PNAS 2009 106 (37) 15583–15587; published ahead of print August 24, 2009, doi:10.1073/pnas.0903620106

Outing, S. (2005). The Blog-Only News Diet. *Poynteronline*. Retrieved June 16, 2005 from www.poynter.org/content/content_print.asp?id=66794andcustom=

Oyama, S., Griffiths, P. E., and Gray, R. D., eds. (2001). *Cycles of Contingency: Developmental Systems and Evolution*. Cambridge, MA: The MIT Press.

Perry, S. and Massie, K. (2007) A Historiographic Look at On-Line Selling: Opening the World of the Private Collection. *Convergence: The International Journal of Research into New Media Technologies*, 13(1), pp. 93–103.

Putnam, R. D. (1995). "Bowling Alone: America's Declining Social Capital," *Journal of Democracy*, V. 6, No.1

_____ (2000). *Bowling Alone*. New York, NY: Simon and Schuster.

Postman, N. (1985). *Amusing Ourselves to Death: Public Discourse in the Age of Show Business*. New York, NY: Penguin.

_____ (1992). *Technopoly: The Surrender of Culture to Technology*. New York, NY: Alfred A. Knopf.

Radway, J. A. (1996). *Reading the Romance: Women, Patriarchy and Popular Literature*. Chapel Hill, NC: University of North Carolina Press.

Rash, W. (1997). *Politics on the Net: Wiring the Political Process*. New York, NY: Freeman.

Resnick, M., and Lergier, R. (2002). *Things you Might not Know About How Real People Search*. Florida International University, Miami. www.searchtools.com/analysis/how-people-search.html.

Rice, R. and Love, B. (1987). Electronic Emotion: Socioemotional Content in a Computer-Mediated Network. *Communication Research, 14,* 85–108.

Robinson, S. (2004). "The Ongoing Search for Efficient Web Search Algorithms." In *SIAM News*, Volume 37, Number 9, November.

Rogers, E. M. (1995). *Diffusion of Innovations*. New York, NY: Free Press.

Russell, B. (1992). *The Basic Writings of Bertrand Russell*. London: Routledge.

Salter, L. (2004). Structure and Forms of Use: A Contribution to Understanding the "Effects" of the Internet on Deliberative Democracy. *Information Communication and Society*, 7(2), pp. 185–206.

Sartre, P. (1956). *Being and Nothingness: An Essay on Phenomenological Ontology*. NewYork, NY: Philosophical Library.

Schafer, R.M. (1977) *The Tuning of the World*. New York, NY: Knopf.

Scheflen, A. E. (1979). On Communication Processes, in Nonverbal Behavior: Applications and Cultural Implications. A Wolfgang, ed. New York, NY: Academic Press.

Schor, J. (1991). *The Overworked American: The Unexpected Decline of Leisure*. New York, NY: Basic Books.

Schudson, M. (2003). "The New Journalism," in *Communication in History: Technology, CultureS society*. David Crowley and Paul Heyer, eds. Boston, MA: Allyn and Bacon (pp. 138–45).

Schutz, A. (1970). *On Phenomenology and Social Relations*. Chicago, IL: University of Chicago Press.

_____. (1972). *The Phenomenology of the Social World*. Chicago, IL: Northwestern University Press

Selwyn, N. (2004). Reconsidering Political and Popular Understandings of the Digital Divide. *New Media and Society*, 6(3), pp. 341–62.

Sennett, R. (1977). *The Fall of Public Man: the Forces Eroding Public Life and Burdening the Modern Psyche with Roles it Cannot Perform*. New York, NY: Knopf.

Shapiro, A. (1992). *We're Number One: Where America Stands—and Falls—in the New World Order*. New York, NY: Vintage Books.

Shefrin, E. (2004). *Lord of the Rings, Star Wars*, and Participatory Fandom: Mapping New Congruencies between the Internet and Media Entertainment Culture. *Critical Studies in Media Communication, 21*(3), pp. 261–81.

Sherman, C. (2002). "Google News Search Leaps Ahead." searchenginewatch.com/searchday/article.php /2160891.

Showalter, E. (1997). *Hystories: Hysterical Epidemics and Modern Culture*. London, UK: Picador.

Sigman, S. (1987). *A Perspective on Social Communication*. Lexington, MA: Lexington Books.

Simonson, P (2006). *Politics, Social Networks, and the History of Mass Communications Research: Rereading Personal Influence.* Thousand Oaks, CA: Sage.

Sperber, D. and Wilson, D. (1996), "Fodor's Frame Problem and Relevance Theory," *Behavioral and Brain Sciences*, vol. 19(3), pp. 530–532.

Sproull, L. and Kiesler, S. (1986). Reducing Social Context Cues: Electronic Mail in Organizational Communication. *Management Science,* 32 (11): pp. 1492–1512.

_____ (1991). *Connections: New Ways of Working in the Networked Organization.* Cambridge, MA: MIT Press.

Sterling, G. (2005). "Yahoo and Google: Man vs. Machine." *Search Engine Journal.* www.searchenginejournal.com/

Stiglitz, J. (2003). *Globalization and its Discontents.* New York, NY: W. W. Norton and Co.

Stone, Rosanne A. (1995*). The War of Desire and Technology at the Close of the Mechanical Age.* Cambridge, MA: MIT Press.

Sundar and Nass (2001). "Conceptualizing Sources in Online News." *Journal of Communication*, 51(1), March, pp. 52–72.

Sunstein, C. (2006). *Infotopia: How Many Minds Produce Knowledge.* Princeton, NJ: Princeton University Press.

_____. (2002). *Republic.com.* Princeton, NJ: Princeton University Press.

Turkle, S. (1984). *The second self Computers and the Human Spirit.* New York, NY: Simon and Schuster.

_____. (1995). *Life on the Screen: Identity in the Age of the Internet.* New York, NY: Simon and Schuster.

_____. (2009). *Simulation and Its Discontents.* Cambridge, MA: MIT Press

Umble, D. Z. (1996). *Holding the Line: The Telephone in Old Order Mennonite and Amish Life.* Baltimore, MD: The Johns Hopkins University Press.

Urry, J. (2007). *Mobilities.* Cambridge, UK: Polity.

Wallis, C., and Stepote, S. (2006). "How to Bring our Schools out of the 20th Century." *Time.* Dec 10.

Walther, J. B. (1994). Anticipated Ongoing Interaction versus Channel Effects on Relational Communication in Computer-Mediated Interaction. *Human Communication Research 40,* pp. 473–501.

_____. (1996). Computer-Mediated Communication: Impersonal, Interpersonal and Yyperpersonal Interaction. *Communication Research, 23,* pp. 3–43.

Weaver, R. (1953). *The Ethics of Rhetoric.* Davis, CA: Hermagoras Press.

Weider, D. L., and Pratt, S. (1990). On Being a Recognizable Indian among Indians. In D. Carbaugh (Ed.) *Cultural Communication and Intercultural Contact.* Albany, NY: SUNY Press.

Werthheimer, L. J. (2003). "Students who Live on Campus Choosing Internet Courses" www.scribd.com/doc/13218240/Students-who-live-on-campus-choosing-Internet-courses

Wicks, R. H. (1992) "Schema Theory and Measurement in Mass Communication Research: Theoretical and Methodological Issues in News Information Processing." *Communication Yearbook 15*, pp. 115–145.

Winner, L. (1977). *Autonomous Technology: Technics-out-of-Control as a Theme in Political Thought.* Cambridge, MA: MIT Press.

Xing, W., and Ghorbani, A. (2004). "Weighted PageRank Algorithm." Paper presented at 2nd Annual Conference on Communication Networks and Services Research (CNSR 2004), 19–21 May 2004, Fredericton, N.B., Canada. IEEE Computer Society. glass.cs.unb.ca/ias/papers/xingw_weighted.pdf.

Zeller, T. (2006). Gaming the Search Engine in a Political Season www.nytimes.com/2006/11/06/business/media/06link.html

Index

About the Author

Robert C. MacDougall is professor of communication and media studies at Curry College in Milton, Massachusetts. His teaching and research center on the cognitive, social, and epistemological roles played by communication media and technology. His publications include *Cultural Technics* (2009, Verlag VDM) a book about the use of email by a group of Mohawk Indians, *Digination: Identity, Organization and Public Life in an Age of Small Digital Devices and Big Digital Domains* (Fairleigh Dickenson Univ. Press, 2011), as well as numerous journal articles and book chapters. His most recent investigations incorporate the use of EEG technology to study the neuro-cognitive effects of screen versus page reading and, more generally, the role played by a variety of multimedia in and for the *brain-body system*. In addition to his teaching and research on media, he is a cyclist, snowboarder, gardener, an avid member of the Media Ecology Association, and father of two willful boys who wish their parents would be willing to watch a great deal more television.